AN ECOLOGICAL AND EVOLUTIONARY ETHIC

DANIEL G. KOZLOVSKY

AN ECOLOGICAL AND EVOLUTIONARY ETHIC

PRENTICE-HALL, INC.
Englewood Cliffs, New Jersey

Library of Congress Cataloging in Publication Data
KOZLOVSKY, DANIEL G
 An ecological and evolutionary ethic.

 Bibliography: p.
 1. Ecology—Addresses, essays, lectures. 2. Evolution—Addresses, essays, lectures. 3. Biology—Philosophy—Addresses, essays, lectures. I. Title.
 DNLM: 1. Ecology. 2. Environment. 3. Evolution.
 Qh541 K88e 1974
 QH541.145.K68   301.31   73-17206
 ISBN   0-13-222943-9
 ISBN   0-13-222935-8 (pbk.)

© 1974 by Prentice-Hall, Inc.
Englewood Cliffs, New Jersey

*All rights reserved.
No part of this book
may be reproduced in any form or by any
means without permission in writing from
the publisher.*

10   9   8   7   6   5   4   3

Printed in the United States of America

PRENTICE-HALL INTERNATIONAL, INC., *London*
PRENTICE-HALL OF AUSTRALIA, PTY. LTD., *Sydney*
PRENTICE-HALL OF CANADA, LTD., *Toronto*
PRENTICE-HALL OF INDIA PRIVATE LIMITED, *New Delhi*
PRENTICE-HALL OF JAPAN, INC., *Tokyo*

ACKNOWLEDGEMENTS

Grateful acknowledgement is made for permission to quote from: G. G. Simpson, *Biology and Man* (New York: Harcourt, Brace and World, Inc., 1969); George C. Williams, *Adaptation and Natural Selection* (Princeton, N. J.: Princeton University Press, 1966); D. Blazek, "The Impossibility And The Wanting", from *I Advance With A Loaded Rose* (San Francisco: Two Windows Press, 1969); Aldo Leopold, *A Sand County Almanac* and *Round River* (New York: Oxford University Press, 1949, 1953); Walt Whitman, *Leaves of Grass* (New York: The Modern Library, 1950); Brooks Atkinson, ed., *Walden and Other Writings of Henry David Thoreau* (New York: The Modern Library, 1950); Odell Shepard, ed., *The Heart of Thoreau's Journals* (New York: Dover Publications, Inc., 1961); Charles Darwin, *On The Origin of Species* (facsimile of the first edition, with an introduction by Ernst Mayr) (Cambridge, Massachusetts: Harvard University Press, 1964); Sir Gavin deBeer, *Charles Darwin* (Garden City, New York: Doubleday and Co., Inc., 1965); Dylan Thomas, *Quite Early One Morning* (New York: New Directions Publishing Corporation 1954; Copyright 1954 by New Direction Publishing Corporation; reprinted by permission of the publisher); Gary Snyder, *Earth Household* (New York: New Directions Publishing Corporation, 1969; Copyright 1967 by Gary Snyder; reprinted by permission of New Directions Publishing Corporation and Jonathan Cape, Ltd.); Ken Kesey, *Sometimes a Great Notion* (New York: Viking Press, 1964; Copyright 1963, 1964 by Ken Kesey; reprinted by permission of Viking Press and The Sterling Lord Agency); Richard Brautigan, *The Pill Versus The Springhill Mine Disaster* (New York: Dell Publishing Company, 1968; A Seymour Lawrence Book/Delacorte Press; reprinted by permission of the publisher); Lynn White, Jr., "The Historical Roots of Our Ecological Crisis," *Science* 155 (3767):1205 (Copyright, 1967, by the American Association for the Advancement of Science). Excerpts from "The Mind as Nature" are reprinted from *The Night Country* by Loren Eiseley, by permission of Charles Scribner's Sons; copyright 1967, by Loren Eiseley. There are quotations from Don Cauble, *Early Morning Death Fragments* (Cold Mountain Publishing Co., 1969), from Richard Krech, "The Hashish Scarab", and from D. R. Wagner's *The Footsteps Of The Returning King That Have Been Lost To Us For Such A Long Time It Seems Like They Never Were And Other Poems* (Sacramento, California: Runcible Spoon Press, 1968). Cover photo of the cephalothorax of the jumping spider *Salticus scenicus* (82x) courtesy of M.D. Oberdorfer and S.D. Carlson and the Scanning Electron Microscope Facility of Russell Laboratories, University of Wisconsin. Photo of the tree frog *Hyla* courtesy of S. Hunsaker.

CONTENTS

Preface, IX

EVOLUTION

Note
1. Origins, 3
2. Evolution, 4
3. Why Versus How, 7
4. Purpose, 10
5. Mind, 12
6. Soul, 13
7. To Whatever Abyss, 14
8. An Evolutionary Ethic, 15

ECOLOGY

9. Noda, 19
10. Lithosphere, Hydrosphere, Atmosphere, Biosphere Noosphere, 20
11. Entities, 21
12. Complexity, 22

Note
 13. Complexity II, 23
 14. Complexity III, 24
 15. Condensed Milk, 26
 16. Goose Music, 28

ANIMISM

 17. Empathy, 33
 18. Noises, Etc., 34
 19. Animal Time, 35
 20. Body Atoms, 36
 21. The Molecular Respect, 38
 22. Owls, 39
 23. Frost, 40
 24. Earth Contact, 41
 25. Biologists, 42
 26. Warfarin, 43
 27. Cocks, 44
 28. The Animal Trip, 45
 29. Scientific Animism, 46

HUMANISM

 30. Context, 51
 31. Meaning, 53
 32. Cultural Lamarckism, 54
 33. The Meaning of Adaptation, 56
 34. Sex, 58
 35. Babies, 60
 36. Woman, Don't Tamper Too Much, 62
 37. Humaning, 63
 38. Boundaries, 65
 39. Freedom, 67
 40. It's Not So Far, 68
 41. Everybody Has Got To Take On The Job At The Top, 69

NATURALISM

 42. What Are You Living For?, 73
 43. What Kind Of Environment Do I Need To Be Me?, 74
 44. Look Where I've Got To Now!, 75

Note

45. It Has To Be Done, 77
46. Don't, 79
47. Alternative Realities, 80
48. The Moment And Forever, 82
49. There Is Nothing Else To Do!, 84
50. Ornithology, 85
51. Diversity, 87
52. Savoring, 88
53. Trips, 90
54. Dope, 92
55. Go!, 94
56. Chock Full Of Stuff To Get You Off, 96
57. I Could Have Stayed At Home Where There Isn't Any Water, 97
58. What Can You Afford?, 98
59. I Need More
60. Do You Lock Your Door?, 100
61. The Pleasures Of Poverty, 101
62. Parameters Of An Acceptable Human Ecology, 104
63. The Fundamental Rule, 106
64. Death And Recycling, 107
65. A Naturalistic Ethic, 109
    Nice Things To Read, 112
    Index, 115

PREFACE

An environmental problem is anything that keeps any person from being healthy and happy. Most of our environmental problems are the results of our own human behavior; it follows that we are not going to solve these problems until we stop behaving in a manner that causes them.

That's not easy; it is hard, very hard indeed, for an individual to change his behavior and seek a more sane and less destructive personal ecology (it means changing many things, including the work that you do, the way that you live, and the *things that you believe*); it is harder still to convince large numbers of people to consider the consequences of their total behavior and modify this behavior so as to alleviate its destructive results. Most of these results are so subtle that it may be nearly impossible for mankind to understand them and stop the process of deterioration.

I maintain that a human society without environmental problems is possible, a society that relates nondestructively to its natural environment and is organized in such a way as to provide each individual with a healthy personal environment; meaning, by the latter, a physical, social, psychological, and philosophical environment that is satisfying and is consistent with what each of us is, what the rest of the world is like, and how the whole got to be that way.

Needless to say, the present culture cannot relate nondestructively to the natural environment; it does not provide very many acceptable personal environments or nondestructive personal relationships, and it has as its philosophical foundation ideas and principles that are so inconsistent with what we know of biology, so superstitious, so wretchedly ignorant and prescientific as to be utterly ridiculous. If the present social structure caters to any aspects of human nature at all, it is to the most base and perverse, to human gullibility, greed, and stupidity.

There are times when I think that the present efforts to deal with environmental problems—the attempts to clean the air and waters, to recycle cans and bottles, to fight power plants and dams and highways—are worse than useless in that they merely prolong the agony. Until there are fundamental changes in the social order, accompanied by equally fundamental changes in our view of man and of reality, such efforts are only stopgap measures. The social tree is diseased because of the decrepitude of its philosophical roots; it will do little good to treat the withering branches.

The roots for a new philosophical system, a view of man and his meaning and his relationships with the rest of the living world, already exist in the knowledge that biologists have carefully acquired in the last century. What is needed now is the social gathering of these roots into a solid philosophical and ideological trunk from which mankind can generate new, beautiful, and environmentally harmonious branches, new ways of life that are personally fulfilling, consistent with what we are and what we need to have to be fully human, and carefully cognizant of how we must relate to each other and to the natural world.

These essays are an outline of some aspects of an *evolutionary and ecological ethic.* We must fashion a philosophical and ethical system that will maintain us within an acceptable human ecology. The present work attempts to provide an evolutionary, an ecological, and a humanistic perspective. We will not have an harmonious human ecology until we understand how we and other living things evolve and are related in the fabric of life that clothes this globe. We will not have a beautiful human world until we acquire a naturalistic philosophy. These notes may be considered an effort toward a responsible animality.

Part I, *Evolution,* is designed to give the reader a comprehensive view of the world of an evolutionary biologist. The biological world view is of considerable importance if we are to understand where we came from and what we are. It is the thesis of these essays that such an understanding is basic to the search for answers to questions concerning how we ought to act in order to manifest the beautiful adaptations of which we are composed and avoid the destruction of the earth in the process. This section is perhaps the most technical and difficult. The reader is advised to read it slowly and carefully; it is important for all that follows.

The second part *Ecology,* is designed to provide the sort of ecological perspective that is sadly lacking in most books about environmental problems. That perspective sees man as an organism that is constantly exchanging materials with the rest of the organic world; man is part of an immensely complex and beautiful web of life which he, in his ignorance, is unraveling and destroying.

Part III, *Animism,* is a collection of notes that emphasize some of the beautiful animal aspects of man. We shall not find satisfactory answers to our environmental questions until we understand what sort of organisms we are. We must learn to accept and savor our capacity for animal empathy and earth contact. Such an awareness, such a desire, is the only escape from an otherwise inevitable human world totally cut off from its origins and ultimately self-destructive. In animism and earth contact lie the preservation of our sanity and, ultimately, a meaningful existence.

But man is more than merely animal, he is *human-animal,* and Part IV, *Humanism,* considers some of the glories and frustrations of that condition. The environment, acting through the evolutionary process, produced us, but we are now drastically changing that environment. We have, through our cultural development, escaped the bilogical evolutionary process, at least in some respects; it is the contention, in this sections that although we may frustrate the evolutionary process now, we cannot escape what it has made us; we must provide ourselves with a context and meaning within the biological framework or destroy ourselves.

The final part, *Naturalism,* explores some possibilities for the development of an alternative philosophy, behavior, and life style designed to bring us into a beautiful, meaningful, and responsible equilibrium relationship with a sustaining earth.

DANIEL G. KOZLOVSKY

EVOLUTION

NOTE 1. ORIGINS

> Rise after rise bow the phantoms behind me
> Afar down I see the first huge Nothing, I know I
>   was even there,
> I waited unseen and always, and slept through the
>   lethargic mist,
> And took my time, and took no hurt from the fetid
>   carbon.
>
> **WALT WHITMAN**

Would we behave differently if we could see ourselves and all life as parts of a natural process in the development of the earth?

We know now that the origin of life was a process that followed inevitably from the bonding capacities of this planet's atoms, that a long period of spontaneous chemical synthesis preceeded it, that molecules sloshed and bonded as their atomic potentialities dictated, that in a rich warm chemical soup at the edge of some primeval sea some of them managed to develop the capacity to direct their own replication; that is, they "got together and became alive."

Life is the inevitable outcome of the gradual evolution of matter; if the right chemicals exist under the right conditions long enough, living molecules result. The process then develops a complexity to assure its survival and reproduction; the thunderous cavalcade of living forms follows.

What we know of the origin of life is a triumph of scientific materialism: no mysterious life force, no supernatural explanations are needed. An understanding of the chemistry of life is sufficient explanation for its origin. If there be any secret of life, it lies in the wonderful bonding capacities of ordinary carbon.

Do you look for a closeness to the earth and its processes? You are nothing but an interesting combination of earth's rocks, water and air; these and two billion years of evolutionary explorations, new trials, new combinations, new forms. Can you imagine the earth-awareness of a society wherein each child is taught that he, like all other forms of life, is a trial in the art of surviving and adapting? All are trials in the art of living, parts of a process the wondrousness of which is exceeded only by the philosophical implications of that inevitable origin and subsequent evolution; for beauty comes in knowing what you are and whence you came and why you be, earth child.

> This then is life,
> Here is what has come to the surface after so many
>   throes and convulsions.
>
> **WALT WHITMAN**

NOTE 2. EVOLUTION

> Before I was born out of my mother generations
>   guided me,
> My embryo has never been torpid, nothing could
>   overlay it.
>
> For it the nebula cohered to an orb,
> The long slow strata piled to rest it on,
> Vast vegetables gave it sustenance,
> Monstrous sauroids transported it in their mouths
>   and deposited it with care.
>
> All forces have been steadily employ'd to complete
>   and delight me
> Now on this spot I stand with my robust soul.
>
> **WALT WHITMAN**

A century ago Charles Darwin shook the biological sciences to their core and provided in his *On The Origin of Species* a revolutionary new system of explanation for the complexity, diversity, and marvelousness of the living world. His fundamental proposition—evolution by natural selection—has been fully substantiated and is the basis of modern biology. Here is Darwin's definition of natural selection:

As many more individuals of each species are born than can possibly survive; and as, consequently, there is a frequently recurring struggle for existence, it follows that any being, if it vary however slightly in any manner profitable to itself, under the complex and sometimes varying conditions of life, will have a better chance of surviving, and thus be naturally selected.

Darwin's great work did three things: it showed how untenable were the then existing theories (creationism and catastrophism); it showed that the theory of descent with modification, already proposed by others, could make sense of the organic world; and it provided a mechanism, *natural selection*, that was able to explain evolutionary change (E. Mayr). The explanation is both simple and powerful; Thomas Henry Huxley, England's greatest zoologist, is reported to have said, when he finished reading the *Origin*: "How extremely stupid not to have thought of that."

Today biologists think in terms of populations and gene pools. We know that mistakes (mutations) in the transmission of the hereditary blueprints, although they provide the raw material for evolutionary change, do not themselves determine the direction of that process. We have come to realize that the individual's genetic message is only an expression of the possibilities of genetic combination within the population; it is the population, interacting with the environment, that composes the genetic message. since some combinations are successful and others are not. To put it simply, the individual organism cannot evolve; it is born, develops, reproduces or not, and dies. It does its damnedest to adapt, but it cannot change its hereditary material. It tries its message against the environment, and if it survives and reproduces, that message is represented to a greater or lesser degree in the next generation. Messages leading to less successful reproduction (numbers alone are not most important; what is crucial is the number of offspring that survive to reproduction), or to no reproduction at all are slowly removed from the population, simply because they are represented in the next generation in a smaller proportion. Contemporary biologists understand natural selection as *differential reproductive success*; the population changes its hereditary material, that is, it evolves, as a result of the selective process.

The economist Kenneth Boulding has suggested that all living forms develop an "image" of their environments; those whose genetics and developmental processes provide a proper image of the surroundings (which are constantly changing, since they are primarily made up of other evolving forms) are those that are adapted and survive to reproduce. One might look upon a gene pool's changing image, its attempts to stay adapted (survive), as its "trip". Thus we can speak of cat gene pool trip ("catness"), orchid trip, maple tree trip, orangutan trip, *Homo sapiens* trip.

Look inside yourself, my friend; the reptile, the fish, the echinoderm, the coelenterate, the slime, and the putrid chemical soup are still there—you and your species code are but a thin veneer of two billion years of trying to save your *image*.

Evolution is the production of new codes of information. A modern biologist is interested in how the existing codes, represented by all the diverse living forms including man, came to be the way they are. I like to poke at my molecular biologist friends that they are studying the biology of the very early Precambrian periods, and that we organismic and extraorganismic biologists study what has happened in the last 500 million or billion or so years.

Life was inevitable—its genes necessarily programmed into themselves the instructions for self-reproduction (those genes that didn't aren't around anymore); living forms then proceeded to more complexity to assure survival of their gene pools, or, in a Darwinian sense, adaptation is a continuous changing of the programming to permit survival, the continuous refining of each gene pool's image of its environment.

But the direction of life was not implicit in its origin, nor in the atomic organization of its genes. Just as the uniqueness of life is in its chemical organization, so is the uniqueness of the multitudinous living forms to be found in the history of the processes that called them into being. *It is the unpredictable historicity of the selective process that provides the interesting diversity of living types.*

The Origin of Species established the fact that the harmony of the living universe is not the wisely planned work of a creator, but the product of the interaction of natural forces (E. Mayr). Darwin did something more; he provided each of us with the possibility of an inexhaustible love affair with the natural world, for the evolutionary perspective leads us to look about us closely, to ask how living things got to be the way they are. We have learned to ask *why* in a new and rewarding way. Why this plant? Why that animal? Why monkeys, why apes, why man, why mind?

> Let others finish specimens, I never finish
>   specimens,
> I start them by exhaustless laws as Nature does,
>   fresh and modern continually.
>
> **WALT WHITMAN**

NOTE 3. WHY VS. HOW

>...and books that told me everything about wasps except why.
>
>**DYLAN THOMAS**

There are about seventy-five thousand species of wasps. There are potter wasps that build clay pots having beautifully formed juglike necks; the wasps provision the pots with caterpillers and lay their eggs within. There are paper wasps that construct elaborate papier-mâché nests of chewed up wood. There are organ-pipe mud daubers that shape their mud nests like organ pipes against a flat stone wall or on boards under a roof. There are thread-waisted wasps, sand-loving wasps, velvet-ant wasps, solitary wasps, digger wasps, thousands of kinds of wasps.

One of the objectives of evolutionary biology is to try to understand how all these species of wasps came to be the way they are. To understand *how* they came to be is to understand *why*.

How did organ-pipe mud daubers come to be? Well, we know that the evolution of predation and nesting in wasps goes something like this; the first and primitive wasps merely found a prey animal, laid an egg in it (the egg would develop into an internal or external parasite and consume the host) and went their way; next, it is hypothesized, there developed the capacity either to drag the prey into some protective niche after egg laying or perhaps merely to shove the prey back into the crevice from which it was extracted; the advantage was obvious—the dead or paralyzed prey animal containing the wasp egg is less likely to be carried off by some other predator (a bird, for instance) or parasitized by other wasps. (There may have been some closure of the niche or cavity.) The next stage in the evolution of wasps consisted of finding a prey, *building a nest* (usually by digging a burrow), laying the egg in or on the prey, and closing or sealing off the nest chamber. Next, the sequence was partly reversed; *a nest was built first,* the prey was sought, the egg laid, and the nest closed. The advantage here was considerable; wasps at the previous stage in the evolution of nest-building behavior must momentarily set down the prey while digging the nest—and robbery of untended prey is very common. Then developed the behavior of building a nest, laying an egg in it, and bringing several prey animals for a food cache; then, provisioning over several days, provisioning of several cells alternately, etc.*

And how did our mud daubers evolve? Well, the Hungarian entomologist

---

*For a technical discussion see H. E. Evans and M. J. W. Eberhard, *The Wasps,* (Ann Arbor: The University of Michigan Press, 1970).

Laszlo Moczar has considered this problem.[†] Earlier workers had hypothesized that mud daubing evolved from an extension of this sequence: use of pellets of mud to close the natural niche of the prey after egg laying; use of more extensive mud application to close a dug nest; partial construction of a nest from mud (to enlarge the cavity); total mud nest development. It is possible that this sequence is correct, that it does represent the stages in the development of mud nest construction. (If you are patient, friends, it will be very clear in a few pages what this has to do with environmental problems.) Moczar, however, noticed that the wasp *Paragymnomerus* (in a family related to that of our organ-pipe mud dauber) is a *water carrier* (instead of a mud carrier) and uses the water to soften the dry clay in the banks where it builds. The resulting mud pellets are *removed* and *dropped* from the nest site. However, early in building, the wasp uses some of the pellets to build a chimneylike turret at the entrance to the burrow; when it is finished with the entire task, the wasp seals off the structure with mud. Moczar therefore postulates that mud-daubing nest construction evolved like this: water carrying to soften the nest substrate; use of some of the resulting mud to seal the entrance; extension or reinforcement of the entrance and subsequent sealing; reversal of the procedure to start carrying mud to serve these purposes; total mud construction. These are two pathways proposed for the evolution of the sophisticated capabilities of mud-daubing wasps.

On the isle of Madagascar lives a small, large-eyed primate, the aye-aye (*Chiromys madagascariensis*), which feeds on beetle grubs it finds beneath the bark of forest trees. Having chewed the bark aside, the aye-aye is faced with the problem of getting the grubs from their burrows in the wood. To do this he uses the middle finger of either hand, which is remarkably adapted to the job at hand (so to speak). For his middle fingers are very long, and more importantly, are very narrow, about one-third the diameter of the other fingers; the zoologist refers to them as "attenuated". Each hand, in other words, has a built-in long, slender, beetle grub extracting probe.

The zoologist's explanation of this magnificent adaptation goes like this: during the evolution of the aye-aye and the development of its preference for beetle grubs, it almost certainly began by eating grubs immediately available beneath the rotten bark of dead and dying trees. Within the aye-aye population, however, were some individuals with variations of finger length and thickness so that with their relatively thin fingers they were able to pry grubs from crevices which the fat-fingered aye-ayes could reach. In times of food shortage or large population size, these narrow-fingered aye-ayes were at a distinct advantage; they waxed fat and prospered, passing to their abundant offspring the genes that produce the thin-fingered variety. With a

[†]See Evans and Eberhard.

continuous selective pressure over thousands of generations, that is, the continuous overrepresentation in the next generation of the genes of the thin-fingered, and therefore successful, individuals, genes for fat and stubby middle fingers simply were selected out, and that magnificent long slender middle finger became the norm. A zoologist would continue his inquiry by asking whether or not such selective pressures are still in operation, or whether other considerations have now negated any advantages of increase in length or thinness.

The point should now be clear. *How* did mud daubers and their organ-pipe nests evolve? Well, wasps that utilized each subsequent stage in the evolution of this behavior were more successful than those that didn't. *Why* did mud daubers and their organ-pipe nests evolve? Well, wasps that utilized each subsequent stage in the evolution of this behavior were more successful than those that didn't. *How* did the attenuated middle finger evolve? Well, there was a selective advantage for those aye-ayes with narrow fingers because they could probe deeper and obtain more food, and they were more successful. *Why* did the attenuated middle finger evolve? Well, there was a selective advantage for those aye-ayes with narrow fingers because they could probe deeper and obtain more food, and they were more successful. And so on.

To an evolutionary biologist, how an organism came to be, and why it came to be, *mean the same thing.*

How did man come to be? Well, if you consult an anthropologist he will tell you what is known of the evolution of the primates and of that anthropoid line that led through the Australopithecines to the hominids to man. *Why* did man come to be? Well, if you consult an anthropologist he will tell you what is known...

All organisms are bundles of adaptations, images that are accurate enough to survive and keep their gene pools going. That's *how* they got here. That's *why* they got here.

> I see that the elementary laws never apologize...
> **WALT WHITMAN**

NOTE 4. PURPOSE

> The wild gander leads his flock through the cool night,
> *Ya-honk* he says, and sounds it down to me like an invitation.
> The pert may suppose it meaningless, but I listening close,
> Find its purpose and place up there toward the wintry sky.
>
> **WALT WHITMAN**

There was a time when the living world was thought to have some direction or purpose, perchance to glorify the creator, or to produce that "highest form" in the "great chain of being" (man), or (and this, of course, is the pernicious concept that we are yet to be rid of; I like to hope that these essays might help do so) that all beings had been placed here on this earth for our enjoyment and sustenance. Today we know better.

True, we understand the purpose of a great many things. We know the reasons for the colors, patterns, and displays of many bird plumages. We know the purpose that flowers serve; those beautiful petals and sweet odors are sexual assists of course, attractants for the insect pollinators that plants with showy flowers must have to help them "get a little." Our capacities to enjoy them are an evolutionary aside, probably related to the fact that our primate ancestors acquired color vision in their search for luscious fruits, which had evolved to attract animals who subsequently disseminated their seeds, that is, deposited them with fertilizer, all over. We understand the purpose of nuts, seeds, sticktights. We can see the purpose of elaborately constructed nests, the purpose of the red slash on a Herring Gull's bill, the purpose of goose bumps on your skin when you are excited or chilled, the purpose of mimicry, of toe reduction in horses, of spider webs, of the honey bee's dance, of the baleen whale's baleen, of bird migration, social behavior, schooling, flocking, a pine tree's resin, a cat's whiskers, a male moth's antennae. We understand the purpose of sexuality; a good biology course should teach you why it feels so good to climb into bed with someone of the opposite sex.

We look for explanation in terms of natural selection, in terms of how these adaptations allowed the species to survive. A century of hard work since Darwin has provided us with an understanding of the purpose of hundreds of thousands, perhaps millions, of adaptations of animals and plants.

There were misunderstandings along the way. Earlier in this century, when biologists were still confused about the genetics of the selective process, there were physicists and chemists who surmised that the evolutionary process had as its direction a trend away from the disordering processes of the universe, an ordering trend against entropy, a small antientropic effort in a degrading universe. A description of the process, perhaps, but hardly its purpose. There are physicists and chemists and biologists without decent evolutionary perspectives who still believe that energy considerations determine the direction of the evolutionary process; but then, there are still people who think the world is flat.

And there are modern heresies. In our rush to defend the balance of nature from human onslaught, ecologists have perhaps been too quick to embrace a theoretical position that treats the natural community as a "superorganism" and subordinates the species and its development to the community's or ecosystem's need for internal balance, diversity, or the like (more later). Don't misunderstand. Within natural communities, within associations of plants and animals, there are many species that have evolved various dependencies and other types of relationships with each other (mimicry and symbiosis are good examples), but that is not the same as maintaining that these relationships *had* to evolve to serve those purposes.

We know today that the evolutionary process has no direction, no overriding scheme, no final goal. The living world is not going anywhere. Except for the process outlined in the previous notes, that is, the procreant urge of each gene pool as it does its damnedest to adapt and survive, evolution has no purpose. *Nor do we.*

>If nothing lay more develop'd the quahaug in its
>callous shell were enough.
>
>**WALT WHITMAN**

NOTE 5. MIND

> (For I know I bear the soul befitting me, I too
> have consciousness, identity,
> And all the rocks and mountains have, and all
> the earth,)...
> **WALT WHITMAN**

Modern neurophysiologists and psychologists work on the tacit assumption that mind is a neural phenomenon. We believe today that mind follows from brain and experience, that it *is* brain and experience. George Gaylord Simpson points out that the old mind—body problem was like the old phlogiston theory of the chemists. As conceptual frameworks mature, these false problems are not exactly solved; they are just forgotten because they become irrelevant.

An evolutionary biologist cannot help but conclude that "mind is a process...that occurs in the body" (Simpson).

Why shouldn't each man have the opportunity to look into the human evolutionary process to try to understand the selective forces that produced this marvelous brain and that provided him with the capacity to think! The biologist G. C. Williams asks: "Is it not reasonable to anticipate that our understanding of the human mind would be aided greatly by knowing the purpose for which it was designed?"

Anthropologists are sure that the answer lies somewhere in the extension of optic and manipulative capacities to language and social behavior. We evolved that mind-brain because we needed it; books, sophisticated culture, philosophical systems are merely decorations superimposed on it, and the capacity to investigate systems of irrational numbers and obscure metaphysical propositions are accidental results of the acquisition of that massive brain.

The scientific and Darwinian revolutions in the social mind mean that in one small instance *the evolutionary process has become conscious of itself.* "Mind" is molecules in an arrangement of such complexity that they are aware of themselves in that arrangement.

It is an immense preciousness.

> Behold, the body includes and is the meaning, the
> main concern, and includes and is the soul...
> **WALT WHITMAN**

NOTE 6. SOUL

> Come, said my soul,
> Such verses for my body let us write
> (for we are one)...
>
> **WALT WHITMAN**

As with mind, with soul. For what is soul but mind and emotions, experience and memory, personality. Each, of course, is unique, for the possibilities inherent in the genetics of sexual recombination are so nearly infinite that each man has his own genotype, and even identical twins experience sufficiently different environments that they develop unique personalities.

Soul is anguish and terror and loyalty and love and perception and doubt and despair and the whole of literature, an attempt to understand ourselves, what we are and can be.

As each soul is unique, so also is each valuable, for each man may be enriched by the perceptions and feelings of many and become more than he would otherwise be (hang on, friends, the ethics are beginning). And there are no limits, other than your capacity to absorb and understand.

We seem to think that a man's soul is eternal in proportion to how well he is known and remembered; thus are invented science and books and teaching and monuments and dynasties. Thus also, arise political and industrial empires and the acquisition of material things; institutions that are not only destructive to the environment and the personality, but which give lie to the Christian theology (since we do not believe in immortality elsewhere we work like hell to find it here). I ask you, why work for it at all; it is the worst sort of an illusion.

Soul is the knowledge that you are unique, it is the opportunity to see and feel and taste and smell the processes by which you came to be *and to wonder that you did*.

Biological man is body, mind, and soul. He is produced as his genotype experiences the environment. When we mess with the environment, we mess with souls. Maybe we ought to be careful.

> And I will make the poems of my body and of mortality,
> For I think I shall then supply myself with the poems
> of my soul and of immortality.
>
> **WALT WHITMAN**

NOTE 7. TO WHATEVER ABYSS

> Stout as a horse, affectionate, haughty, electrical,
> I and this mystery here we stand.
> WALT WHITMAN

I steal the title from that of a section of Homer W. Smith's *Man and His Gods*, wherein Smith, having carefully traced the origin of our religious ways, concludes, as any evolutionary biologist must, that all our ideas of gods and devils follow from our ignorance and fear, and that it is high time to get over that nonsense, accept what we know, quit fearing what we do not understand, and learn to follow the path of intelligent inquiry wherever it may lead, to whatever abyss.

We know now that life was inherent in the chemicals and conditions of this earth. We understand the selective process by which the diversity of forms, ourselves included, came to be. We must accept, as a condition of our existence, that the cumulative accidents and purposive adaptations of that historical process produced us, bodies, minds, and souls. It turns out, as we now realize and as Alan Watts in his book, *The Wisdon of Insecurity*, beautifully expressed it, that the rocks of this earth were "peopling rocks" after all. It did not have to be that way, but it was; and we are.

One of the reasons advanced for the correctness of a belief in a God or gods is that all human groups have had such beliefs; as if the earth must be flat because all men believed it so, as they once did and were wrong. In the past all human groups were ignorant and scared; many people, unfortunately, still are ignorant and frightened.

The capacity to imagine gods and devils is a stage in the development of the human mind, a stage compounded of ability and ignorance. Interestingly enough, the mind image of God can be now seen as a capacity of unenlightened brain, and as such it is to be considered not as something responsible for the evolutionary process, but as a product thereof. Like many other products of the process this one is now defunct, having been superseded by the better capability of rational inquiry, and must be gotten rid of to maintain our adaptive stance.

It is time to learn to stand on our own two feet, to see what we can be and what we can understand, to fear nothing but our own ignorance and stupidity. G. G. Simpson in his book, *Biology and Man*, states, "One hundred years without Darwin are enough."

> ...And I say to any man or woman, let your
> soul stand cool and composed before a million
> universes.
> WALT WHITMAN

NOTE 8. AN EVOLUTIONARY ETHIC

> Shall I postpone my acceptance and realization
> and scream at my eyes?...
> **WALT WHITMAN**

It should be clear by now that the evolutionary process is an entirely nonethical one, a process entirely indifferent to its results, a mechanism of continuous species adaptation without consideration of bad or good, a phenomenon that produces immense numbers of parasites and diseases as well as men and orchids. Measured against the enormity of geological time, survival, reproduction, and continuous adaptive change are the only abilities of any value.

On the face of it, then, it is obvious that there are no ethical principles to be derived from a consideration of the evolutionary process. To speak of the right to life of sandhill cranes or pasque flowers or passenger pigeons or bald eagles, or man for that matter, is to overextend an ethical system already wobbling on decaying foundations.

But an evolutionary perspective can supply some principles helpful to the generation of a naturalistic ethic. *It is good to remain adapted*; the alternative is extinction. *It is essential for the full development of any organism that it encounter an environment to which it is adapted*, for the organism is a bundle of adaptations; in an inappropriate environment these adaptations are frustrated and the organism becomes distorted or diseased. And *it is essential that a species not destroy the environment that it needs*.

We must fashion an ethical system consistent with this perspective; that is, a system that allows us to remain adaptive, that provides each human individual with the environment he needs for full animal and human development, and that does not destroy the surface of the earth; *we must abandon ethical systems that cannot see man in this evolutionary and ecological context*.

Total intellectual and emotional acceptance of our evolutionary origins has some important ramifications. Since all living forms have essentially similar basic biochemical systems (the Krebs citric acid cycle of respiratory importance is common to both plants and animals, meaning that it is so fundamental that it evolved before plants and animals went their separate ways), and since they clearly have as much meaning and purpose as yourself, it follows that to demean or devalue them is to demean and devalue man; *if you are to respect the wonderousness of yourself, you must respect other living organisms as equally purposive, equally beautiful adaptive systems*.

It is ironic that in our efforts to make our personal lives acceptable we cheapen them with our fundamental disregard for other living things. This cheap aggrandizement of ourselves, this wretched disrespect for the rest of the organic world, follows inevitably from those contemptible religious ideas of the corruptness of man; when we come to accept ourselves for what we are, as perfect as we have managed to be for the conditions to which we have been exposed, no more, no less, and complete and meaningful enough in our organic selves, only then will we discover that the greatest human dignity follows from a respectfulness of everything that is as meaningful as ourselves, that is, the entire living world.

The evolutionary perspective allows us to revere ourselves and all living things as beautifully adapted systems, magnificent arrangements of earth stuff.

> I believe that a leaf of grass is no less than the
>   journeywork of the stars,
> And a pismire is equally perfect, and a grain of sand,
>   and the egg of a wren,
> And the tree-toad is a chef-d'oeuvre for the highest,
> And the running blackberry would adorn the parlors of
>   heaven,
> And the narrowest hinge in my hand puts to scorn all
>   machinery,
> And the cow crunching with depress'd head surpasses
>   any statue,
> And a mouse is miracle enough to stagger sextillions
>   of infidels.

WALT WHITMAN

ECOLOGY

NOTE 9. NODA

> Strange and hard the paradox I give,
> Objects gross and the unseen soul are one.
> **WALT WHITMAN**

Sometimes I think that my boy puss, Rat Face, is really just a device cleverly designed to change cat food into cat dung. I know, of course, that this is wrong, although there is no question about his abilities in this respect. The biologist that I am recognizes that the boy puss is really something else, considerably less efficient, and that is a living device for changing cat food into cat!

A living organism is a system that continually draws materials and energy from its environment into itself. This material and the energy locked in its chemical bonds is rearranged, restructured, and ultimately incorporated into the living protoplasm. Eventually it is released from this organization, either excreted bit by bit, or freed by decomposers after the organism's death. In an absolutely fundamental way, therefore, an organism is a system for taking selected parts of the environment and organizing them into its own being, a system for rearranging them into its own essence, *into its own image of how the world ought to be.*

There are some important lessons here. It becomes obvious that the atoms of this planet are not randomly distributed in its rocks, water, and air; rather these ancient and primeval associations are covered by a thin film of living organization. Organisms are "nodes," or lumps, or hot spots of organization in the undifferentiated sea of environment, fantastically arranged collections of earth elements, some of them associations of such complexity and intensity that they can run or fly or sing or purr or think!

It also becomes clear that the organism-environment dichotomy is nonsense; all living things are open systems, constantly exchanging atoms with other systems, living and nonliving—there are no impermeable boundaries. Don't be misled by that thick skin of yours; when you cease to exchange, eat, excrete, breathe, you're dead.

The organism-environment comparison is fundamentally meaningless. The atoms are transients; what is environment today is organism tomorrow, what is organism tomorrow is environment the next.

We should be careful of what we dump into the environment, because, physically and psychologically, tomorrow it's likely to be us.

> ...O soul, we have positively appear'd—that
> is enough.
> **WALT WHITMAN**

NOTE 10. LITHOSPHERE, HYDROSPHERE, ATOMOSPHERE, BIOSPHERE, NOOSPHERE

> I find that I incorporate gneiss, coal, long-threaded moss, fruits, grains, esculent roots,
> And am stucco'd with quadrupeds and birds all over,
> And have distanced what is behind me for good reasons,
> But call any thing back again when I desire it.
>
> **WALT WHITMAN**

The earth began most probably as a cloud of stellar dust, condensed itself to solid rock (the lithosphere), and condensed upon it water (hydrosphere), leaving some gaseous elements represented in an atmosphere. When enough of the right constitutents of the lithosphere and atmosphere were dissolved in the hydrosphere, life (biosphere) originated. One of the most interesting aspects of the biosphere is the evolution of a whole new human mental sphere (noosphere) that envelops the earth and, we vainly think, the universe. Hydrosphere and atmosphere interact with lithosphere. Elements of all three are incorporated into biosphere, which in turn affects each of the others. About 20 per cent of the existing atmosphere, for instance, is oxygen, almost all of which has been placed there by living plants; most of the iron ore of the famous Mesabi Range of Minnesota was deposited by bacteria, etc.

Biosphere is a condensation, a reorganization of lithosphere, hydrosphere, atmosphere; *it cannot be separated from them,* it cannot do without them, since *it is a form of them,* constantly exchanging materials with the others. Noosphere is a marvelous expression of a new development in the arrangement of part of the biosphere, and is therefore a rearrangement of the other spheres. Your thoughts derive from the interaction of rocks, water, and air in the neurons of your brain! By this means can rocks, water, and air think and contemplate themselves in this corner of the universe.

NOTE 11. ENTITIES

> The poet does not need to see how meadows are something else than earth, grass, and water, but how they are thus much. He does not need discover that potato blows are as beautiful as violets, as the farmer thinks, but only how good potato blows are.
>
> H. D. THOREAU

There is a danger, however, in developing the capacity to see organisms as their gene pools' new trials or as enveloping noda in a soup-like environmental matrix. The danger lies in comprehending the phenomenon in such a way as to miss its concrete representation; it is possible to forget to see the organism. It is possible to forget to see humans as individuals, too.

My friends, Ginny and Stan, are fascinated by aquatic invertebrates. They are likely, whilst traveling around, to stop at any inviting stream, get out their nets and screens and pans and bottles, and "sample" it for animal life. Now they know damn good and well that they are not thus engaged in any *scientific* endeavor, although I doubt that they would admit that; they are interested in what is in the stream, I suspect, not as part of some large abstract scheme of zoogeography, but because they are interested in animals as entities, they want to see and know these individual animals, here, now. This is the delightful disease of the mind that bird watchers get too, when they get beyond the "bird list" stage.

Ginny and Stan have that wonderful ability that naturalists have always had: the capacity (indeed, necessity) to interject themselves into the living fabric, to put themselves into the midst of natural phenomena and to say, consciously or unconsciously, "I'm going to try to see what goes on here; and I want to try to see it raw, the way it is, not as my human abstraction."

The type of biologist I can understand is one who looks at an organism and subconsciously, or sometimes, when he is particularly exuberant or a little snookered, consciously says, "Hello, animal" or "Hello, plant," not meaning "Hello, representative of this species or that," not "Hello, representative of this type or that, this family or that, this order of phylum or that," not "Hello, reminder of this evolutionary pathway or that, this remarkable life style or that, this magnificent illustration of what the living material can do or has done," not "Hello, organism responsible for this behavior, this role in the assembled community of organisms;" but who, although he is aware of all of this (how else could we label him biologist?), still has the capacity to see the wondrously functioning individual, *the node as concrete momentary reality*, as real and permanent as himself, the plant or animal as entity who says, "Hello, old friend of my garden or pasture or hedgerow or forest or seashore; and how are *you* today?"

NOTE 12. COMPLEXITY

> The press of my foot to the earth springs a
> hundred affections,
> They scorn the best I can do to relate them.
>
> **WALT WHITMAN**

Ecology (from the Greek *oikos* = household; thus, the study of the household relationships of the living world) is usually defined as the study of the relationships of organisms to their environments, or the study of environmental biology, or scientific natural history, or the study of the economy of nature, or of external physiology. Some wags hold that it is a discussion of what everybody knows in language that nobody can understand, or an elaboration of the obvious, or that branch of biology abandoned to terminology. At any rate, it is an attitude rather than discipline, an attitude that looks for meaningful phenomena outside of the internal chemistry or physiology of the individual organism. Ecologists who are also evolutionary biologists (many are not; their ecological perspective accordingly suffers an unfortunate simplicity) have the additional pleasure of observing natural phenomena in the context beautifully expressed by G. E. Hutchinson as *The Ecological Theatre and the Evolutionary Play;* they can see the unfolding of the evolutionary process in the theatre of the natural world.

The complexity of the organic world is enormous and naturally frustrates the inquiring mind. So enormous is the number of organisms, so infinite their many and varied relationships, so subtle and far-reaching are the influences of their constant activity and evolving natures, as to laughingly belittle those persons who, usually lacking an extensive familiarity with the phenomenon, think that they will manufacture simple models of the system to facilitate their understanding. Physicists badger the fundamental material of the universe in their attempts to understand the complexities of interaction of some fifty "elementary particles," an enormous task. Would you care to try to understand the possible interactions and relationships to other living things of 300,000 species of beetles? It is against this horrible complexity of the living world, appreciated by anyone familar with it, that the tremendous synthesis and explanation achieved by Darwin's great mind is to be measured.

Not only is the natural world complex beyond imagination, it also eludes understanding (not, I hope, appreciation) because it is constantly changing; the ecologist enters this shifting matrix at his own risk. Some delineate small aspects of the ecological theatre; others carefully investigate small parts of the evolutionary play. No man can encompass the entire performance. A man can, of course, try; I shall maintain later that this is one of the few genuinely valuable things to do with your life.

NOTE 13. COMPLEXITY II

What rolls, laughs.
SAMUEL SAMAHLIN

There was a time when I was young and foolish enough to try to outwit old Dame Nature, when I tried to simplify and abstract from her principles of community organization and metabolism that would ease the burden of comprehension. Slowly but surely the simplifying vision was destroyed on the hard rocks of observation and experience. I remember well an important crack in my naive systematizing. It was on a warm summer night, as I sat reading next to an open window, that the old Dame smiled a bit and showed me something of her unexpected and inexhaustible content. For my naked arm, underneath the lamp, was covered with tiny insects, nearly microscopic little bees or wasps (Hymenoptera) that were no larger than the periods on this page that you now read. My attention was drawn from the dead and dreary abstractions of my zoological texts to an incredibly minute and beautiful living organism.

Ah, little Hymenopteran, I said to myself, you are a magnificent thing. Why hasn't my zoological education prepared me to understand you? Is it true that in that little body of yours, as small as a period, there is a nervous system rapidly firing, unraveling the signals from your magnificently beautiful microscopic eyes and antennae and commanding your six legs and four fragile veined wings? Is it true that you are covered by little sensitive hairs and that the molecules of oxygen swirling around my arm are, together with my pipe smoke, really being moved into those little spiricles and air channels that permeate your tiny body, there to be incorporated into your very living stuff? Is it true that you and your thousands of fellows have been brewing, generation after generation, in some flowers, or damp cells, or dung heaps, going through immensely complex metamorphoses, taking the very materials of the earth into your complex and sure systems, and generating such complexity and order and beauty without me knowing it? Is it true that in this immeasurably complex and huge world you now dare to seek and will find a mate and will experience ecstasies only known or dreamed of by the two of you? Is it true that in that warm moist evening, in that dark green world outside my window, in that soil and air aswarm with insects and other living forms, is it true that your microscopic life is as beautiful and meaningful and important as mine?

Why doesn't it tell me that in my venerable tomes; or do I look in the wrong place for the wrong thing?

NOTE 14. COMPLEXITY III

> Color, which is the poet's wealth, is so expensive
> that most take to mere outline or pencil sketches and
> become men of science.
> H. D. THOREAU

Evolutionary biologists are interested in gene pools; we used to call them species and some of us, of course, still do. There was a time, not so very long ago, when we thought of organisms as individuals, and though we were aware that some were alike (and reproduced together), others different, we gave little concern to the significance of these similarities or differences other than to thank the Creator for his wisdom and generosity. Linnaeus changed all this, and there followed two hundred years of systematizing; activity so intense and productive that it appeared, not long ago at all, we would someday finish the task of placing organisms in their proper species categories; we would understand nature, in that sense, completely. A modern biologist now asks himself this question: "If I had as much information as I needed, could I place any organism into the proper species designation?" The answer, interestingly enough, is *no*.

It turns out that the Old Bitch is difficult, and the more we learn about gene pools the more impossible it is to carefully delineate them. The natural world will not stand still, and those genes keep escaping into combinations where they do not belong. It turns out, in other words, that whilst there are a large number of groups of organisms whose gene pools are easily encompassed, who only breed with their recognizable kind ("good species"), there is an even greater assemblage among animals, and particularly among plants, where the composition of interbreeding groups is so complex, so indefinite, so rapidly changing, that it laughingly defies categorization. Are the two populations now diverging? Have they become sufficiently distinct so that if they came back together they could not breed? Or would they interbreed only slightly, with limited success? Is this hybrid swarm a step in the direction of gene pool separation (speciation), or is it an example of convergence (introgression)? What the hell *is* going on?

The more we investigate, the more difficult and complex the problem becomes. Soon we realize that we need new definitions of species to encompass our new found gene pool arrangements, and a number are rushed to the fore. But we learn more, and even these become inadequate—we experience a constant struggle to keep our system of classification, our system of scientific understanding, sufficiently sophisticated to keep up with our observations. And we soon have a dilemma. For if our system of explanation becomes so refined as to properly include all the variables, all the observed gene pool arrangements and aberrations thereof, we soon have no system at all, merely a description of the observables. Science is most productive when it is ignorant.

The natural world is so complex that we cannot even delineate its constituents. Can we be so naive as to think that we will understand its operation, that is the multitudinous and almost totally unknown relationships of these ill-defined constituents, by some simple categorization or systems analysis?

Education is the process of hybridization in the human mind.

NOTE 15. CONDENSED MILK

> ...and the cost of a thing is the amount of what I will call life which is required to be exchanged for it, immediately or in the long run.
>
> H. D. THOREAU

There was a man I nearly knew once upon a time in Alaska (he was recently dead, I can't tell you how; I lived for a summer in a cabin that he had built, in a cabin within sight of his own, his cabin being tended by the homesteaders down the road who used his garden plot and kept their crock of sauerkraut in his kitchen, whose daughter made us bread and crazy cakes, and who themselves didn't know who was supposed to take care of old Frank's things, and so they did, because around those parts neighbors are responsible for a man's things; and one day when I was in his cabin to get some sauerkraut dipped by handfuls out of the slime-covered crock—it wasn't the best I've ever had, but we ate it and were glad to have it—that day the homesteaders showed me the sharpened nail on the kitchen wall where Frank had opened the cans by puncturing them), who, I was told, drank canned condensed milk—no fresh milk—for all the years he lived there, forty-some, I disremember exactly. You should have seen the pile of cans.

I had this thing about condensed milk; I used to keep some in the house, although I'm not exactly sure why. It makes the best cup of cocoa in the world, but otherwise I don't like the taste of the stuff. It's a security thing, like all our environmental problems; I want to have the milk available but not to have to worry about it going sour. And when it's gone, I have the can.

It is indicative of our shortsighted approach to environmental problems that we think that the answer to cans and bottles and such is to recycle them. The answer is to not use them unless they can be washed and refilled. To recycle them we need to mine the materials and acquire the energy to make and run the machines to crush and fragment and melt and remake them, again and again, in a diabolical effort to wear the atoms out, a process ultimately self-defeating, for if only a small percentage is not recycled but is irretrievably disseminated across the landscape, the process is diminished by that amount, and time after time, cycle after cycle, all the material is eventually lost.

With a great deal less effort, with a great deal less destructive environmental impact, each human individual could be furnished with all the permanent, refillable containers that he could possible need. When we went from bucket of beer to returnable bottle there may have been no very far-reaching changes, *except for the disastrous consequences of the diminishment of neighborhood,* but when we went from returnable to throwaway, and now to recyclable, we fundamentally changed our relationship with the earth. The demon convenience has us.

(Jack King can maybe tell you about the rhubarb wine we kept in the cellar of that cabin; the old-timer down the road gave it to us so that someone would drink it because he *needed the bottles* for the next batch. Old Axtel once got so drunk that he stayed drunk for three days, after he shot a bear through his cabin door; not through the doorway, *through the door.*)

Don't believe this "evaporated" milk nonsense, friends; the *water* is evaporated, the milk is *condensed*.

NOTE 16. GOOSE MUSIC

> I will go to the bank by the wood and become
> undisguised and naked,
> I am mad for it to be in contact with me.
>
> **WALT WHITMAN**

I stand, each summer, in a prairie that produces its nostalgic fullness of native grasses and forbs, of big bluestem grass and *Silphium* sunflowers, both taller than a man, in the undisturbed tiny strip of land between a highway and a railroad track. I have hardened myself to the sounds of roaring automobiles and thundering trucks; I accept the inglorious fact that unless I wish to drive a good long way (itself destructive to my search for some aesthetic pleasure from the natural world) to find a prairie that does not have the intruding sounds of its mechanical conquerors—*if there by any such prairie with solitude at all—I had better accept the noise.* At least for prairie I will act as though a remnant of its natural beauty were better than no taste at all. But I cannot so easily accept the blank and uncomprehending faces in the speeding automobiles that pass so close to me, faces without interest or imagination, faces oblivious to everything about them, unable to recognize or appreciate the remnants of the magnificent flora that once covered a third of the continent and gave it soils that were four feet thick, unable to see them even as "weeds," faces (I swear it's true) unable even to see *me* there by the roadside. What must such a life be like? And where do you suppose are they all going so fast, when pieces of the primal beauty of the vast American prairie lie here to be seen and touched and smelled in the roadside ditch?

Watching the fullness of prairie bloomings, or watching the spring migration of birds, each of these is like developing a love affair, it gets better and better, each gets more intriguing, each is infinite in its possibilities for intimacy, affection, and pleasure. The natural world has an unlimited capacity to add to the breadth of the human experience. You have an obligation to learn how to approach your primal lover.

There is an immense amount of the energy and materials of this earth locked up in the multitudinously fascinating bird forms of organization. Hundreds of millions of these astounding organisms twice yearly hurtle themselves across the face of the globe. Millions of geese *ya-honk* and cackle their ancient ways north and south, bringing spring to the tundras and the tundras to the human soul. I ask myself, sometimes, if the shifting masses of their fragile bodies might not make a detectable tilt in the axis of the earth; I suppose not, but there is no doubt that they can, and should, make a detectable shift in the capacities of the human mind. Would you be so careless as to miss it? Would you be so stupid as to destroy it (to destroy part of yourself)?

> And when the dawn-wind stirs through the ancient Cottonwoods, and the gray light steals down from the hills over the old river sliding softly past its wide brown sandbars—what if there be no more goose music?
>
> **ALDO LEOPOLD**

ANIMISM

NOTE 17. EMPATHY

> By destroying pagan animism, Christianity made it possible to exploit nature in a mood of indifference to the feelings of natural objects.
>
> **LYNN WHITE, Jr.**

My friend Mendelson has a pet raccoon that has the run of his house; it terrorizes his cats (except old Ug), playfully bites his children, eats anything it can get its hands on, defecates in a cat box, and feels most secure in the basement or in Jon's lap. The coon is fascinating to watch as she explores her surroundings. For racoons, you see, have a surprising intelligence and a very delicate, gentle, manipulative ability; to watch her seek out objects, handle them, puzzle over them, and play with them is an incredible lesson in zoology. To have her exercise her gentle manipulations on you, to see her genuine pleasure when you scratch her back, these are exercises in the unlimited capacity for empathy with animals.

On my way home late one night, around a sharp curve in the road, I came upon five young coons the same size as Jon's—a couple of them in my path—swerved sharply and just missed them! Almost left one there, a bloody corpse strewn across the road. And the loss? Well, what about the loss? Hundreds of coons are run down every day. And then slowly the mind remembers that beautiful, fascinating coon at Jon's, intelligent, curious, delicate, friendly, *and the mind realizes that they are all that way.*

NOTE 18. NOISES, ETC.

When we hear that sound of the crickets in the sod,
The world is not so much with us.

H. D. THOREAU

And so a note about what is wrong with buildings and motors and constant noises; but an essay in animal empathy too—*empathy with yourself* and what you need to be healthy and happy.

Fat Jon works in a windowless lab at the top of a huge building near the compressors or whatever, where there is a constant low motor noise. It bugs him. But old Dame Nature smiles on those who look at her closely, and somewhere near Jon's lab something, probably a rat, died, and a fecund blowfly found it, and one day when Jon came to work his lab was aswarm with noisy buzzing blowflies. There was nothing to be heard but the peace and tranquility of animal sound.

Took my friend Skinny John a bucketful of cabbage seedlings and left them in the unfathomable dreariness and deadening grayness and printed-page dryness of his office in a huge institutional building. He forgot them there; fortunately they had a lot of water. They also evidently had eggs of cabbage butterflies upon them, so one day, a couple of weeks later, he opened his office door to discover hundreds of butterflies fluttering about. (I must remember to get my bucket back.)

Mr. Soleri, can you build me a single-building city where 250,000 people will live and work, where they will hear or feel no motors? Where blowflies will find the dead rats; where butterflies might happen? Where I can hear my owls?

NOTE 19. ANIMAL TIME

> My foothold is tenon'd and mortis'd in granite
> I laugh at what you call dissolution,
> And I know the amplitude of time.
>
> **WALT WHITMAN**

Warm summer night, Oberdorfer and I sitting slightly inebriated on my lawn (snookered, I suppose, because we're tired of our cares resting from lives too busy and responsible, from days hurrying to finish never-ending tasks); from the rim of my valley, swelling and fainting through the freshly leaved and deep-breathing woods, comes a strange new sound: *clip-clop, clip-clop, clip-clop, clip-clop, clip-clop, clip-clop, clip-clop.*
"What the hell is that?" says he.
"Shhh. Listen."
Clip-clop, clip-clop, clip-clop, clip-clop.
"Is it some kind of *animal*?"
"Just listen, damn it."
Clip-clop, clip-clop, clip-clop, slow and rhythmic, loud now and disappearing, quiet and sober and comfortable and reassuring.
"What the hell *is* it?"
"I guess its one of my Amish neighbors, going home."
"This time of night?"
"What's wrong with this time of night?"
"Well, nothing, I guess. There's a man got plenty of time."

> Logic and sermons never convince,
> The damp of the night drives deeper into my soul.
>
> **WALT WHITMAN**

NOTE 20. BODY ATOMS

Where have you disposed of their carcasses?
Those drunkards and gluttons of so many generations?
Where have you drawn off all the foul liquid and meat?
I do not see any of it upon you to-day, or perhaps
    I am deceiv'd,
I will run a furrow with my plough, I will press my
    spade through the sod and turn it up underneath,
I am sure I shall expose some of the foul meat.

WALT WHITMAN

When you come to accept the fact that your body, your mind, and your soul are all organic entities, when you come to the realization that you too are an open system that is constantly exchanging materials with your environment, taking these into your living being, then you begin to pay attention to what you eat. Physical, mental, spiritual well-being follows from a healthy diet. Thus is it that the organic food "nuts" are careful about what they eat, being concerned that it should be whole, unprocessed, unpoisoned. There is a widespread belief among organic gardeners—correct. I think—that chemical fertilization leads to an imbalance of mineral nutrients in the soil, subsequently to a less than perfectly healthy plant, subsequently to a less than adequately nourished animal. There is a widespread belief—correct, I think—that contemporary commercial processing is destructive to the vitamins, proteins, amino acids, and probably hundreds of other important compounds in whole fresh food; a destruction that is only partially compensated for by "enrichment." The biologist that I am finds it incredible that anyone with even a rudimentary knowledge of the extremely sophisticated cellular chemistry and the fragility of the absolutely essential enzymatic systems, could think of taking into his own body substances known to be poisonous and used, therefore, as preservatives because they retard oxidation or prevent any bacteria from utilizing the food before we get to it.

Do you know of some way to weigh and measure the convenience of deteriorated, processed, poisoned food against twenty or thirty years of ill-health, or a loss of a decade or so of your personal existence? It seems to me that the assumptions of Christianity are responsible for the fundamental disrespect that we show our bodies; I shall maintain elsewhere that only Christians and other psychopaths can be in a hurry to suffer and die.

Do you know where your body atoms come from? Do you care? Can you feel when and where they were grain, trees, horses, other humans, eagle, bison, rodents, worms? Thus the friendliness of manure and the need to accept ourselves to the extent of recycling our own defecations.

When you eat a banana you eat atoms that were once tropical vegetation, before that tropical hummingbirds, before that nectar in tropical flowers, before that giant lianas or tropical trees, before that an Indian's defecation, before that a milpa field's corn seeds, before that rotten jungle humus, before that monkey, before that lush red fruit, before that orchid, before that—ever since those atoms were formed in the heart of some unimaginably ancient sum.

You have made these cycles too, if you can know, understand and imagine it. Would you ignore this, or forget this?

I suggest that a psyche that is in complete harmony with itself and the entire world would want to acquire its atoms from worlds it comprehended, would want *to see itself coming*, would want itself to know the paths of its origins, would want to prepare its own new molecules, would want to know which plants it came from, which fruits, which defecation, which soil, which calling owl, which rotting leaves, which piece of earth. *Psyche and body can determine what will become psyche and body, and wallow in the deliciousness of the process.*

Had an old yellow cat, Fart-boy, once upon a time, and when he lost the capacity to maintain his personal organization, I buried him in my garden and planted one of my favorite herbs above him. Now when I use that herb (it grew lush and beautiful), his atoms become part of my body and my brain, and in a true and fundamental sense, *I have my Fart-boy back again.*

> We are the lights we think we see
> coming at us from behind.
> If we step back to let us pass,
> we'll get by just fine.
>
> **JOHN INGHAM**

NOTE 21. THE MOLECULAR RESPECT

> One farmer says to me, "You cannot live on vegetable food solely, for it furnishes nothing to make bones with;" and so he religiously devotes a part of his day to supplying his system with the raw material of bones; walking all the while he talks behind his oxen, which with vegetable-made bones, jerk him and his lumbering plow along in spite of every obstacle.
>
> H. D. THOREAU

We need a culture that respects and admires atoms and all their wondrous combinations. We need a culture that recognizes the beauty in living organization, that understands the fundamental life process of utilizing the sun's energy to build complex molecules from simple earth stuff, that can revere biological organization wherever it is observed, that can recognize and maintain natural processes in all their marvelous complexity, that understands the nature of the recycling processes of this good earth.

Good food, good health, the good life comes from good soil; soil that has incorporated into it biological matter undergoing decay: humus, compost, manure, call it what you will—soil thirsts for broken bits of once living organization, food for its own multitudinous organisms whose actions are the only way to furnish growing plants with all they need. The earth's soils (we mistakenly think of them as "our own") are thirsty for organic matter, thirsty for the water that only decaying humus can collect and hold. The earth's soils are starved because we continually exhaust them and refuse to return an equal amount of nutrients in a healthy complement of organic matter.

It has recently been proposed that we take animal and human wastes, now collecting in huge stinking amounts in feedlots and cities (there is no fundamental difference) and use them to generate fuel for our automobiles; in short, burn them. This is an immense obscenity, an immense disregard of what the earth needs.

Manure is the most precious substance in the world.

Sometimes I catch myself about to throw some crumbs or onion peels or a leftover bean into the trash that is to be burned, instead of into the slop bucket that goes on the compost heap. There is something immoral about burning plant or animal material that, as compost, could feed the soil; there is something monstrous about such a disregard for the wondrously complex and essential material that even *dead* life-stuff is.

NOTE 22. OWLS

> The wood thrush is a more modern philosopher
> than Plato and Aristotle. They are now a dogma,
> but he preaches the doctrine of this hour.
>
> **H. D. THOREAU**

I like to think that I am sufficiently in tune with the natural world around me that I hear every owl that calls within hearing distance of my house, day or night. Pissing outdoors is essential to this awareness; the invention of indoor plumbing was a monstrous step in the reduction of human awareness of natural phenomena, in the ability to recognize ourselves as earth animals.

NOTE 23. FROST

> We have thus far exhausted trillions of winters
>   and summers,
> There are trillions ahead, and trillions ahead
>   of them.
> **WALT WHITMAN**

"Why don't we dig one up, Stanley?"
"How can we? The ground is frozen."
"There is no frost in the ground, Stanley."
"Wow. I guess I'm out of contact with the earth."

NOTE 24. EARTH CONTACT

The lovers bed was the sole place to enact the dances and ritual dramas that link primitive people to their geology and the Milky Way.

GARY SNYDER

Maybe tomorrow it will rain and I can stay inside and read.

NOTE 25. BIOLOGISTS

> The greatest and saddest defect is not credulity,
> but our habitual forgetfulness that our science is
> ignorant.
>
> **H. D. THOREAU**

Oh, he's all right as a biologist in the lab or lecture hall, but he's not worth a damn in the field.

Or in bed (they say).

NOTE 26. WARFARIN

Warfarin is a chemical that causes internal hemorrhaging in mammals and is used extensively as a rat killer. It is cheap, quick, silent, painless (to you), and until recently, effective (a resistant strain of rats has already evolved); and because the rats crawl to water while bleeding to death internally, they usually leave your house or barn or whatever, and there is no objectionable odor of dead rats. The most important aspect of this poison (most pernicious, I'd say) is that you don't have to see the dead animal at all; you don't have to see it or smell it or handle it, and thus the appeal; the damn rats just disappear.

I suspect that the healthy human person who has to handle the animals he kills begins to do it halfheartedly. For in seeing and touching and disposing of them there is necessarily a communication of some empathy, there is a recognition that you are destroying other living beings. Warfarin cheats us of the genuine human capacity to see and feel ourselves as killers and to have to come to grips with our murderous activities.

There was a time when we had the patience and humanity to try other solutions, like the domestication of cats for instance, and whoever ignores the immense pleasure and benefit that we have received from our association with these fascinating animals does so at the risk of his own intellectual and emotional impoverishment. Today we build houses within the hollow walls of which the rats can elude the most capable cat, and the cat is overfed and made to defecate in a catbox, and loses the meaning of catness, becomes, in fact, as purposeless and as out of context as ourselves.

And this, like thousands of others, is an example of the essentially dehumanizing effects of the application of chemical solutions to pest problems. They are used because they do not allow our humanity to interfer. *We are the poorer for it.*

NOTE 27. COCKS

> One of the penalties of an ecological education
> is that one lives alone in a world of wounds.
>
> **ALDO LEOPOLD**

A cock crows under my bedroom window, a wild jungle call. We never bred it out, probably because it doesn't affect laying capacity or meatiness. That crowing seems as wild and free and primitive as that of the northern Indian jungle fowl from which our domestic chickens are derived. Jungle under my window!

A thousand human generations woke to that wild sound. Philosophies, arts, sciences, were nurtured in days encompassed by crowing cocks. Who will determine the next benefit to the human race of those lusty calls?

The hardest nut to crack of all the difficult nuts of environmental deterioration is the very real human capacity to forget something not now present that was once of considerable importance to our lives, and the obvious inability to miss something we've never experienced. And so from generation to generation the environment becomes less interesting, less diverse, with smaller unexpected content, and our immediate surroundings become depauperate of animals and plants and exuberant human life. What your father can hardly remember, you will not miss. What you now take for granted, or what is now slowly disappearing, your children, not having known, cannot lament.

For a thousand generations men woke to a crowing cock, had interjected into their subconscious lives that small part of jungle wildness. In a couple of generations we have excluded these cocks from town and barnyard, and even the farmer now wakes to an alarm clock. The human world is the poorer for it, but does not know that, *cannot* know that.

Can you regain what you do not know you should have?

NOTE 28. THE ANIMAL TRIP

>The scent of these arm-pits aroma finer than prayer.
>
>**WALT WHITMAN**

If you learn to watch your own animal behavior, you will find an inexhaustible source of fascinating pleasure. Watch your animality; understand and appreciate your animal trip.

After I had abandoned my "bathroom door" investigations as being no longer productive, I began to consider the question of "which side of the bed do you sleep on?" It turns out, interestingly enough, that almost everyone who sleeps with someone else has his or her own side. It is subject to change, especially when moving from one house to another. I've changed with nearly every move, usually to be put on the side handiest to the chores that need to be done, checking the stove or furnace, letting the cats in and out, or with me occupying the side of the bed most open to assault, the side toward the entrance to the room. This is, of course, consistent with the convenience factor, but when I lived in a house with doors to the outside in every room—I would build no other way, it is a wholly beautiful way to relate outdoors to in, every room ought to communicate directly with the outside; I will maintain elsewhere that windowless rooms are fundamentally destructive to the human psyche—in that beautiful old stone house I slept next to the outside door, on the *inconvenient* side of the bed, although now that I think of it, when the weather was not bitter cold, it was handy for urinating off the porch next to the huge clump of lilac and for letting the damn cats in and out.

The real fun comes with changing sides. For one reason or another you sleep on the other's side. Then back to your own side. Then the other's side again. Then, somehow, the distinction fades, and for a short time either side is acceptable. Then the habit sets in again, territory is established, the complaint, "get over on your own side," returns.

What side of the bed do you sleep on? And when is the last time you switched? The same sort of question applies to partners and houses, too. And where do you put your pants when you take them off?

House, room, burrow; you can't appreciate your need for a burrow until you've slept for a few years in a single room with thirty other men. Burrows, entrances. My house, for instance, has summer and winter entrances; in summer, access is directly into the kitchen (it has seven doorways), in winter I use a hallway, having blocked off the kitchen doors, and then I have a cloaca, a chamber that insulates me from the outside world.

You can learn to trip on nondestructive changes, to derive pleasure from changes in your personal animal activities related to changing seasons.

Watch yourself. If you are in an interesting, genuinely stimulating and demanding environment, you are a fascinating *animal*.

NOTE 29. SCIENTIFIC ANIMISM

To confront night, storms, hunger, ridicule,
accidents, rebuffs, as the trees and animals do.

WALT WHITMAN

A scientific animism would be that inquiry which carefully and appreciatively considers ' the animal aspects of ourselves, our capacities to understand other living things and generate an empathy with them; the whole not with an eye to changing our animistic predelications, as though they ill-befitted such cultured beings as ourselves, but rather to discover and emphasize them, to allow us to savor them, to provide a foundation for a genuinely sound human animal ecology, a comfortable human psychology, a reasonable human philosophy.

A scientific animism would consider the fear and the beauty that is to be discovered when you are frightened by a polyphemus moth when it snaps its wings open to display their foreboding eyespots. It would be concerned with understanding all our hopes and fears.

A scientific animism would consider the relations of man to vegetation, of man and his attachment to trees, of his reluctance to settle the prairies and plains, his reluctance to leave the protective forest edge. It would consider, too, fears and awe of dense primeval forest.

A scientific animism would consider the need for sacred groves, the need for green landscape, wildness, quiet, naturalness, openness of spirit, the *need to touched*.

A scientific animism would consider the phylogenetic possibilities of animal empathy (you can *stroke* a mammal), not ignoring a lady I know whose pet Great Horned Owl understands her moods, is happy when she is happy, is down when she is down, and who (I use *who* instead of *which;* you would too if you knew the owl) tries to cheer her up by getting into some ridiculous mischief when she's unhappy.

A scientific animism would consider the multitudinous ways in which animals train *you*.

A scientific animism would consider your territoriality, your aggressiveness, your personal relationships, your drives, your ego, what makes you happy, sad, relaxed, tense. It might discover that only a part of your personality is to be regarded as intellectual, economic, or cultural.

A scientific animism would consider the pleasures and pains of the human animal trip; why it feels so good to be a human, why it is so magnificent to be alive. Such an animistic approach might provide a basis against which to compare the abominable social or cultural situations in which people no longer feel the wondrousness and beauty and animal excitement of living.

A scientific animism would not only consider why it feels so good to make love, but, forgetting the analysis would, like any genuine science, lose itself in the gloriousness of the phenomenon that it seeks to explain. Science functions best when we understand so perfectly that we no longer need it.

> I think I could turn, and live with animals, they
>   are so placid and self-contain'd,
> I stand and look at them long and long.
> They do not sweat and whine about their condition,
> They do not lie awake in the dark and weep for their sins,
> They do not make me sick discussing their duty to God,
> Not one is dissatisfied, not one is demented with the
>   mania of owning things,
> Not one kneels to another, nor to his kind that lived
>   thousands of years ago,
> Not one is respectable or unhappy over the whole earth.
>
> **WALT WHITMAN**

HUMANISM

NOTE 30. CONTEXT

> The conditions of existence of an organism include the conditions of existence of all its ancestors as well.
>
> W. F. CANNON

I have already suggested that the organism is an expression of the potentialities in the gene pool; it has a genotype. This genotype is a phylogenetic memory of what has been successful in the past environments. The selective process means that the new combinations of each generation are faced with this question: which of us is still suitable for the "now" environment? The raw genotype itself, of course, is not operated upon, unless it has some gross deformity; it must be fleshed out, so to speak, and the developmental process does just that. The genotype interacts with the environment to produce the visible organism, the phenotype. There is some leeway in the process; the environment modifies the instructions of the DNA in the genotype, but there are limits to this modifiability. Some aspects of the genotype are expressed in one environment, others are expressed in different environments. There may be aspects of the genotype frustrated for lack of any opportunity for expression. Obviously, the criterion of satisfactoriness, the measure of completeness of environment and subsequent genotype expression, is that the organism should successfully function in the selective context; that is, the organism acquires the morphology and behavior leading to reproductive success. A goodly number of organisms are not reproductively successful; one might consider them to be possessed of the wrong genotype for the environment in which they find themselves, or that they are in the wrong environment for which their genes are programmed (both these views are, of course, the same thing; they are another illustration of the meaninglessness of the organism-environment dichotomy). In either case, in both senses, the organism is *out of context*.

Nothing is as ridiculous, as pathetic, as obscene as an organism out of context.

To be a meaningful, real, interesting entity then, an organism must experience an appropriate environment and undergo the selective process of measuring its adaptations, its capabilities, against that environment. Having done so, it must reproduce to assure the continued existence of genes that work. Reproductive success is, as already suggested, the measure of what works. An organism that finds itself in an unsuitable environment or fails in the process of matching its capabilities against the appropriate species environment fails to reproduce, and, from the viewpoint of the evolutionary process, it is forgotten as uninteresting. The organism out of context, unable to find an environment against which to test its potentialities for success, epitomizes the abyss of meaninglessness.

Listened to a conversation the other night about a lady who kept a parrot for sixty years. Kept the damn thing on a perch in her living room for *sixty years!* (I have seen parrots, wild and free and meaningful, in their own sane way, in the mountains of Panama—sweeping in a thunderous hail of wingbeats and calls from fruit-bearing tree to tree, green and yellow and blue hailstones loose in a moist green jungle of a complexity and beauty unimaginable by those who have not seen it.) For *sixty years,* the parrot was maintained alive *out of context.* Of course it went mad. And the lady?

And that is what is so obscene about zoos and aquaria and caged birds: they are all being maintained *out of context.* Morphology and physiology are maintained without reference to the environments and selective forces that produced them. We see the animals, beautiful as they live and breathe, but we miss the most important lesson; we are deprived of contact with the conditions that produced them, we are denied the opportunity to understand *why the organisms are the way they are.*

Nothing is as abjectly worthless as an organism denied reproductive contact with its population in the selective context, *denied the possibility of finding its meaning in its biological meaning,* placed out of context.

Are you in context? And what does that mean?

NOTE 31. MEANING

> Think of an essay on human life, through all
> which was heard the note of the huckleberry-bird
> still ringing, as here it rings ceaselessly.
>
> H. D. THOREAU

I have suggested elsewhere that it is safe to say that until now reproductive success, sex and its results, has been the fundamental meaning of all living entities. The 'meaning' of existence has been to try your genes against the environment and see if they were good enough to be passed along. 'Meaning' was in terms of keeping the population adapted to a changing environment by generational selection of still favorable gene combinations. 'Meaning' was to try your reproductive damndest.

And the question now is this: where shall we look for meaning in a social, cultural world, a world horribly plagued with overpopulation, a world that measures adaptation not in terms of reproductive success but, instead, changes the environment to encompass all the genotypes, a world that culturally and rapidly changes the entire species environment? Where is there meaning in human culture? If I cannot contribute to continuous fitness of this species by reproducing or not, then what can I do? What is my meaning? What is human meaning? What is meaning in an organism-adaptation-environment system gone mad, a social organism that is determined to change its environments to suit itself and is changing them so rapidly that no genetic correspondence can be hoped for.

Are there people who really still believe that success in living has to do with maximizing the number of their offspring? Do they think they are keeping us adapted? Can they really think it will make any difference to the ways and the satisfactoriness of adaptation to have their *genes* more abundantly present in a world where ideas are changing that adaptational environment minute by minute?

You cannot love, suck, grow, compete, and proliferate fast enough for what we are doing to the system *within* which you have to do the loving, sucking, growing, competing, proliferating.

At least during this momentary phase of human development, this explosive, minding, cultural, populous, planet-manipulating phase, *genetics is dead as a way of finding meaning*. It may catch up again, when things get tough.

NOTE 32. CULTURAL LAMARCKISM

"Have a nice trip, now."
    Receptionist at fancy private lounge at
    O'Hare Airport
"And you too, madam, on your headlong rush to nowhere."
JAMES LOUIS SAWYER

Darwinism is now established for the organic world. Jean Baptiste Lamarck preceded Darwin, and in his *Philosophie Zoologique* (1809) provided a theory of evolution, an explanation in terms of *acquired characteristics*. Lamarck was so far ahead of his time that he was not even dignified by a refutation. We know now that the inheritance of acquired characteristics is quite impossible as an explanation for biological evolution; Lamarck has assumed the oblivion accorded to men who see a phenomenon, but who provide the wrong explanation for it.

But in the human world, in our social world where biological evolution has been superseded by something called cultural evolution, that is, adaptation by cultural means instead of genetic, in this human world evolution is essentially Lamarckian in nature. For what is culture but sets of acquired characteristics, habits, and beliefs, ways of tilling the soil, building dwellings, and engaging in economic enterprises?

Perhaps there was a time when the cultural traits (acquired) were also adaptive, that is, were tested, generation after generation, against the environment and found not wanting. But contemporary cultural evolution consists of the invention of new environments, or in other words, the actual manipulation of the environment to suit our own wishes or convenience. And so what may have been an early imposition of some cultural behavior on a process still essentially biological has become a total predominance of the cultural, with the result that the biological adaptation, genetic selection from generation to generation as a means of population change, is inoperative; whatever genetic changes might take place, such as increase in the frequency of genes related to diabetes, poor eyesight, narrow pelvic canals, etc., are now totally nonadaptive.

And cultural evolution is a self-accelerating process; once power and energy were concentrated in the developing urban centers, these centers had the capacity to continue to draw on the energy resources of the countryside while dreaming up new schemes, new realities, new cultural excesses. In time this process of urban development produced immense populations psychologically and intellectually cut off from their base, cancers capable of destroying the healthy body they are parasitizing. And since the human environment is primarily the cultural environment, *is* the mental environment, and is self-contained in its view of reality, it continues to snowball out of hand; always the human imagination dreams new worlds, new social existences, new madnesses. So far we have been able to absorb these acquired characteristics, the corpse we feed upon is not yet exhausted, the tyranny of human domination and stupidity is not yet complete. It is a positive feedback system nourished by its own reckless exuberance, growing bigger and more dizzyingly demanding of the unlimited capacity of the human imagination. And the majority of living humans perceives no other reality than "progress" and "new" and "unlimited" and "hang on, here we go!"

And there seem to be no rules to this Lamarckian evolution, no particular criteria for the generation of new possibilities and their acceptance— although we know one rule in the economic sphere: an activity is engaged in if it makes someone secure by making him a bundle.

In a very real sense the capacity for cultural evolution is unlimited, since it is constrained, at least in theory, only by the infinite dimensions of the human mind. There are, of course, two checks in the biological "real" world: the self-accelerating, self-generating, self-responsible cultural system eventually comes up against the solid wall of maintaining its life support systems on the surface of this earth, of maintaining a sufficiently "real" notion of human ecology, so that we may continue this sunburst of ingenuity, exploration, madness, if you so wish to call it. And whatever the exuberances of the human mind, wherever they lead us, we must have environments satisfactory for each of the diverse living human natures; in short, we must not strangle our human and animal natures in the convolutions and evolutions of our minds and culture.

It appears that both of these checking conditions are being exceeded. There can be no assurance that they will come to bear slowly and with increasing severity in what an ecologist would call a density dependent manner, thereby righting the situation; there is every reason to believe that these conditions can be (and may have been) exceeded; and if this is so the situation can deteriorate very rapidly.

The important question is whether this freewheeling system of Lamarckian-like evolution will eventually respond with sufficient checks and balances to keep itself adapted in the sense of not destroying its base or producing environments to which we are psychologically and physically illfitted. The answer to this question is unknown.

Such checks can be brought into play in two general ways; by means of the Four Horsemen of the Apocalypse: war, famine, pestilence, death, and their associated misery and madness; or by intelligent reorganization of our world view and behavior so as to maintain the surface of the earth in a condition that will provide for us indefinitely. I opt for the latter; I think that such an understanding is the function of all intelligent inquiry. Understand, please, that I do not advocate all-encompassing manipulation of the world to make it livable for us— rather, the development of the capacity for intelligent *nonmanipulation*. I fear, sometimes, that for the major part of this civilization it may be too late.

NOTE 33. THE MEANING OF ADAPTATION

So the present, utterly form'd, impell'd
by the past.

WALT WHITMAN

In the Darwinian sense, adaptation for the individual is in terms of reproductive success. For populations it is in these terms also, because the population as a gene pool is the totality of these individual successes. When we speak of a population remaining adapted, we mean that it does so automatically as each individual tests his genotype against the environment. In cultural evolution, essentially Lamarckian in nature, there then comes the question of what is adaptation, what kind of human behavior is adaptive, and adaptive to what?

Darwinian adaptation in terms of reproductive success is immediately rejected. Darwinian evolution takes thousands or tens of thousands of generations, and we cannot assist in a general social adaptation by reproduction; our intellectual or economic activities have much more immediate and perhaps far-reaching impact. In short, the enviroments we are culturally providing for ourselves are changing so rapidly that fundamental changes take place within single lifespans, making adaptation in terms of reproduction completely irrelevant.

Nor is it obvious that economic superiority or success is socially adaptive; there is every reason to believe that important social evolution proceeded from the deliberations of men ill-fitted in the struggle for economic gain; poets, philosophers, prophets, inventors, scientists. There is no reason to suggest that such noneconomic cultural change is necessarily adaptive either; witness the destructive effects of the prevailing religious ideas for instance.

The complete title of Darwin's great work is *On the Origin of Species by means of Natural Selection or the Preservation of Favored Races in the Struggle for Life*. In the context of the present discussion, the question is this: is there, within the social milieu, an analagous process of selection for the preservation of favored cultural innovations, and in what sense does such a selection and preservation result in adaptation?

And thus, we come to the unfortunate crux of the situation. For it appears that in contemporary Western society there is a selective process and it preserves cultural changes *in terms of the wealth and power that they produce for some individuals*; this selection is not only nonadaptive, it is definitely maladaptive, since it is *destructive to other individuals and disastrous to the adaptation of the human group to a satisfactory earth environment.*

Unless you wish to "go back to the trees," which in any case you are ill-equipped to do, personal adaptation in the cultural sense means using some of the plasticity of personal adjustment to get along in your social milieu, or finding another, or making your own subculture, hoping all along that the demands of participation in any of these do not exceed some limits of comfortability in personal relationships; that these demands do not, in fact, force you to be something other than you might otherwise be. This is the realm of "culture against man," and any reasonably intelligent and sensitive person has probably had considerable personal experience in the social limitation of his possibilities.

Questions regarding the adaptiveness of the behavior of the social group have been profoundly ignored. It is profitable and necessary to ignore them, because much of the economic activity of this culture is, as Aldo Leopold put it, "pathogenic in respect to harmony with land."

It is obvious that we need to develop a new selective process, a psychosocial selection that is an intelligent means of remaining adapted. And it should also be obvious by now that we speak not of genetic adaptation, but of adaptation in two different but related senses: *individual adaptation, meaning the provision of environments within which each individual can be fully human and happy,* that is, providing him with an environment where his genes are not of context, not inappropriately exposed, and *a social adaptation devised to bring the human species into some sustaining equilibrium relationship with a nondeteriorating natural system.*

These, I submit, are the definitions of cultural adaptation. Anything less is a failure to recognize our potentialities and necessities. Anything less will result in subhuman individuals teasing out their fantasies in a deteriorating world.

We must use our minds to discover what we are, animal and human in each body, so that we can live in environments that are healthy for *both. That is the environmental problem.*

NOTE 34. SEX

> Have you ever loved the body of a woman?
> Have you ever loved the body of a man?
> Do you not see that these are exactly the same to
>   all in all nations and times all over the earth?
>
> **WALT WHITMAN**

Organisms pass their genes to their offspring. It is safe to say that until the advent of sociality and culture, this was the fundamental meaning of each organism. Very early in the history of life, this transmittal took the simple form of more or less equal division of genetic material and cellular contents into two new daughter cells. This meant that the only source of variation within the population, the only handle upon which the selective process could get a grip, was mutation, the change in some portion of the genetic stuff. If an organism has genes XY, then in the absence of mutation, all its offspring had genes XY, too.

Somewhere, a billion or two years back, a new process developed, a process involving a doubling of the genetic complement (*diploidy*) and a reassortment and recombination of its elements in a two-organism cooperative genetic exchange; in short, *sex* was born. If organisms with double genetic complements of XXYY and xxyy reproduce the old way (asexually), all their offspring will be XXYY or xxyy. If, however, they exchange half of their genetic material (sexually), then all of their offspring will be XxYy, that is, different from either parent. And now the interesting part: when two of these offspring themselves reproduce sexually, the next generation has all of these genetic combinations: XXYY, XXYy, XxYY, XxYy, XXyy, Xxyy, xxYY, xxYy, and xxyy; nine possibilities present themselves. Since, as a matter of fact, each organism has thousands of genes, the result of sexual recombination is an almost unlimited variability, an almost unlimited number of new possibilities each generation, an almost unlimited opportunity for the selective process.

Some plants still reproduce only asexually. Other plants and some animals may reproduce both ways. But for all animal species the decision in favor of sexuality was made a billion years ago or so and has become the inflexible rule. It was a decision with important implications; in the context of gene pool continuance, in terms that we have seen are the only ones with any real meaning, till the present, the individual organism became insufficient. It now took two to tangle. It now took two to make a generationally viable unit, a whole meaningful life. It was a fundamental change in the living process. *It was a decision that you cannot escape.*

In the billion years and hundreds of thousands of generations that have followed, the sexual process has become elaborated and so incorporated into anatomy, into physiology, and into the psychology of all sexual organisms that it has become one of the primary aspects of any species' biology. "The vagina is an organ specialized for the reception of the penis; the penis is an organ specialized for transmission of sperm to the vagina." The continuance of the gene pool, reproduction, has been the meaning of life. All other systems in all organisms, in all the history of life, have been subservient to this fundamental purpose; otherwise they, too, were lost. In sexually reproducing organisms, therefore, it is safe to say that all other aspects of their biology, all other systems and abilities, all aspects of anatomy, physiology, and behavior were tested, ultimately against one criterion only: did they lead to increased success in reproduction? Throughout the history of the vertebrates, in all the history of the mammals, in all the sophisticated development of the primates, in every aspect of the existence of proto-man, *sexual success was the aim of existence*, was the meaning of behavior, was the purpose of the elaboration of body and psyche. *Do you think that you can escape it now?* Do you think that you can deny what you are? Do you think that you can escape *yourself*?

The *self* is one-half of the reproductive and psychic *Self*.

> the only cure is human cure
> & one person does not make a human...
> **DON CAUBLE**

NOTE 35. BABIES

> Urge and urge and urge,
> Always the procreant urge of the world.
>
> **WALT WHITMAN**

There are three and one-half billion of us. In thirty-seven years there will be seven billion of us; that increase is damn near inevitable. If we continue the way we are going, in an additional thirty or thirty-five years there will be fifteen billion of us. It has been estimated that the long-range carrying capacity of the earth to support humans in an environment of the sort that I've been trying to outline here, an environment within which each human could develop his total capabilities and live a healthy and happy animal and human existence in a nondeteriorating world, the earth's capacity for this kind of human equilibrium population, this sort of humanity, is about one-half billion individuals; that is about one-seventh of what we already have! To anyone who has considered the problem and its ramifications, it becomes clear that we are in for a very bloody mess indeed. I do not speak of holocaust, though that too is possible, but tremble instead at the apparently inevitable "brave new world" that will result; a rigidly controlled society of billions of *Homo-post-sapiens* (Hugh Iltis). *What we shall become will be determined by how we live, which is inexorably connected to how many of us there be.*

A fundamental principle of animal ecology is that in an equilibrium population without emigration or immigration, births must equal deaths. Obviously population regulation, the equilibrium of births and deaths, can be achieved in two different ways; the birth rate can be decreased to match the death rate, or the death rate can be increased to match the birth rate. "You pays your money and you takes your choice." For plenty of obvious and sound reasons we have made the decision to minimize the death rate and have been modestly successful in that respect. If an equilibrium population is desired (we ecologists see it as not only desirable, but as absolutely essential; the only argument is how it is to be achieved, and at what level) then we must follow the logic of our principles; *having lowered the death rate, we must lower the birth rate to match it.* If we export death control in the forms of food, agricultural assistance, and medicine, it is obvious that we have a moral obligation to accompany that death control with *equally effective birth control.* There are 240,000 humans born each twenty-four hours; about 50,000 die of starvation, disease, and old age, leaving a daily net of about 190,000 persons. Each time another human is added to the population, our so-called inalienable rights are necessarily a little more circumscribed; each time another human screams his bloody way into this good world, our lives necessarily becomes a little more closely controlled. Controlled!

I was a fourth child; the morality I espouse would have flushed me down the toilet.

Why so many babies? We want them. We are stupid or careless. There are no wholly satisfactory birth control techniques, or we have some ridiculous social or religious superstitious beliefs that prevent us from using what is available. We instinctively recognize that for our psychological wholeness, male and female, we must have the experience of generating new human beings. It is a central aspect, perhaps the overriding concern of our animality. It cannot be denied without horrendous psychological consequences. (Until recently the ultimate irony of foregoing the experience of having children, for the female in this culture, was to have to spend your entire life rearing other people's children as a teacher or nun.) *The ultimate lovemaking is that without possibility of contraception.* It is "proof of manhood." Having children is the only truly acceptable expression of the "earth mother" capabilities of each human female. It is the only satisfactory use of those beautiful milk sacs. And it is an immense pleasure to watch a child grow up, to watch an almost undifferentiated blob of protoplasm become, in a few short years, a verbally communicative human person. And an extremely important "trip" can be taken with the realization that two partners have committed themselves to a lengthy common enterprise.

The question is this: can this absolutely essential human need be satisfied by the generation of only one or two offspring? If the answer is yes, our problem is not so difficult, although it is still immense. If the answer is no, then we must find cultural ways; familial or social organizations that can satisfy these needs while holding the births to two per couple or a little less. We will need a long period of negative population growth to reach a long-term equilibrium population.

This is the dilemma of human biology: we must intelligently restrict our heretofore indiscriminate reproduction *without destroying our psyches,* for if we are unable to do so, we will inevitably generate environments also destructive to those psyches. "You pays your money and you takes your choice."

> Oh women fit for conception I start bigger and nimbler babes.
> (This day I am jetting the stuff of far more arrogant republics.)

WALT WHITMAN

NOTE 36. WOMAN, DON'T TAMPER TOO MUCH

Woman, don't tamper with your body too much until you have had kids. I worry about the hormonal and long-term changes of the pill, and that womb irritated by that IUD is the chamber that must house your children for nine precise months; I worry. Should you make love only with someone whose kid you'd bear if the "damn thing broke" or "whatever" didn't work?

The saddest biological obscenity would be to want to have children and not be able to because you messed up your system with years of preventing children. The saddest, most fundamental, permanent irony.

We need sex, we need mates, we need kids, to be fully human; and there are too damn many of us already. And forty percent of all living humans are pre-reproductive. PRE-reproductive! The one really good solution: after two, the *old man's* tubes are tied. Then neither should be risking physiological or psychological damage.

These remarks may seem odd in a collection of essays such as these, from a biologist who thinks that the ramifications of the population explosion have not been fully considered even by the alarmists—it's going to be a lot worse than even they can see, it's going to be a bad one, and many things will likely go, or almost go: property, democracy, your "inalienable" rights, the mind itself may need straightjacketing, the soul, too, may need to be destroyed, and *1984, Brave New World,* and *The Wanting Seed* will seem like child's play compared to what it may come to before we stop it, before we stop ourselves. Funny for an ecologist to put encouragement to risky business here, in this book, but there is here another thesis, and that is that we cannot give the lie to our bodies, to our minds, to our souls—to be whole and healthy *we must be what we are*, and for most of us that means sex, pair-bond, kids.

Old man, I don't see why it should be an affrontery for me to suggest that you get your damn tubes tied after two; should your child die, there is considerable success these days at having them untied, so you could rework that generational magic. It is more than magic, it has till now been fundamental meaning.

The question is whether we can get smart enough to outsmart what we are, for we are organisms like any other, *designed to let go reproductively* and breed ourselves into oblivion; all species are.

To outsmart ourselves, but not to negate what we are in the process, not to detract from the fullest measure of humanity. Can we be both animal and man, and outsmarted animal and still whole man? If not, we are still only animal, and cannot help ourselves.

NOTE 37. HUMANING

> The notion that the truth can be sought in books is still widely prevalent, and the present dearth of illiterate men constitutes a serious menace to the advancement of knowledge.
>
> **ANONYMOUS**

We have become mind-beasts. I have said something about cultural Lamarckism, meaning that we now adapt to our environment, or rather adapt it to us by cultural, not genetic means, a process which involves the social inheritance of acquired characteristics; thus it is reminiscent of Lamarck's theory of organic evolution. Within that framework is it possible to consider some of the effects of our mental preoccupations, our attempts to adapt to and by a "think world?" It is just possible that this mind-world, this culture-world, might be at odds with our biological needs; I suggest the following consideration.

What we have learned to do with our minds, lately, is inimical to child-rearing, since in the mind-world we wish to continually change communication partners and therefore cannot provide a stable communications environment for the child, a situation which is destructive for him. The child has to change his communications relationships as fast as the agile minds of his parents; he cannot keep up and becomes confused and disturbed. Because adults who are genuinely interested in changing communications, changing which individuals they are "into" from time to time, are less likely (recognizing this "minding" characteristic in themselves) to settle down and have children, the process may be essentially self-regulatory. Maybe it's possible to conceive that these minding tendencies are ultimately the most important factor in population control, with reproduction being reduced in two ways: by the general madness that total minding might lead to, or by a reduced birth rate that might be the result of a preoccupation with mind communications rather than body or generative communication.

Watching television, reading, thinking out of the family context, preoccupation with the cares of the world, etc., draw us away from communication with our children, with each other. This, I submit, is one of the main causes of all the unhappy people we have running about; they are likely to come from situations of minimum communication because of active, imaginative minds that were "off elsewhere" instead of communicating with the developing child.

And in this sense is our minding, our humaning, dehumanizing, in that it results in the destruction of our children or whomever we might beautifully communicate with in a more basic human-animal way.

The test of the persuasiveness of this contention is whether or not you set this book down *now* to go communicate with your children or your wife (husband, sweetheart, etc.) and find out what they are thinking, and touch them, *right now*.

Well, did you?

(We should have books with fifteen blank pages preceded by "take time to communicate with someone personally;" story takes up fifteen pages later.)

(Then books with fifteen blank pages, and the story takes up *where it would have been if the pages were there and were essential to the story*.)

(Then with thirty blank pages, etc.)

(Then mostly blank, with a beginning and end to the story.)

(Then only a few page beginnings.)

(Then all blank, only a title page and cover.)

(Then a cover only.)

(Then a cover with no writing on it.) (A bookseller's nightmare.)

(*Then no book.*)

NOTE 38. BOUNDARIES

> A woman waits for me, she contains all,
>   nothing is lacking,
> Yet all were lacking if sex were lacking, or
>   if the moisture of the right man were lacking.
>
> **WALT WHITMAN**

We are all busy generating boundaries and relationships. Most of us are conditioned to feel that we need to make some distinction between ourselves and the rest of the world—a distinction that I have already suggested is not well founded in a geochemical view of reality—a distinction which, carried to extremes, isolates us into essentially nonfunctional and meaningless entities. We have our territories, our little securities and insecurities, our privacies. Mine!

Personal meaning can be found in a couple of directions. Meaning might come from relationships: whom do I know in Washington, who is indebted to me, how many exciting new things (or women) am I on top of, how many intense intimate friendships can I carry on simultaneously, how numerous and extensive are the dimensions of *me*? Or meaning might come from within, from a harmony of spirit and nature, from a serenity in the realization of the meaning of participation in the life processes of the earth.

The most difficult thing, of course, is to balance these two, to find some satisfactory relationship between the outreaching into the human community, which is inexhaustible and ultimately dissipating, and the inward seeking for assurance and satisfaction.

The human psyche has been evolutionally tuned to the need for gregariousness, to other-relatedness. In a primitive and stable culture this led to understanding and acceptance, to ego control as well as fulfillment, to identification without dissipation. In the inexhaustible possibilities of complex modern societies this aspect of human nature must ultimately be frustrated, other-relatedness can never be sufficient, ego generation becomes a means of extension instead of being controlled by the process, and the inner self is overwhelmed by insecurities (and is forgotten to ease the pain of the personal meaninglessness that complex societies generate.)

Our environmental problems will not have satisfactory solutions until these elementary questions of personal extension and nonegoistic inner development have some answers, some balance, in a reinforcing social organization. Our personal environmental problems will not be answered until we abandon the impersonality of this complex civilization and make some return to *tribe*. Thus the correctness of the instinctive generation of communal living among disaffected youths—the failures of which are perhaps attributable to the immense difficulty of overcoming early reinforcement of ego extension as a means of identification and security acquisition.

Our civilization has led to an ignorance of ourselves as animals and humans in an evolutionary and ecological context. Not understanding that life comes and goes, not understanding the source of our food or of our meaning, out of context, out of contact with natural processes, out of a livable, genuinely human, social world, we have mistakenly come to the conclusion that our skins are the boundaries to ourselves. It should be clear from what has been said of the importance of sexuality and all its reinforcing anatomy and psychology to the evolutionary process that even in a modern society where reproduction can no longer be considered fundamental to meaning, no man can find the boundary of his personality within himself. There is one fundamental extension of the personality required in the search for psychic wholeness, one existence beyond skin, that must be encompassed. The fundamental conclusion of our sexuality is that a single whole psyche can only be found in two bodies.

It is possible that both the outer and the inner constant searching are processes of sublimation in the absence of a personality fulfillment in total sexual and psychic union. They may also be self-reinforcing aberrancies. The search for meaning in inexhaustible outward relationships has the inevitable result of diminishing the personal sexual relationship, leading to a dissatisfaction that seeks to find fulfillment only in further outreaching, further desperate encompassing of new ideas, knowledge, minds, books, wealth, bodies, leading to still further deterioration of the personal two-person psychic union. And there is no end.

All meaning is encompassed in a single woman (man).

The prevalent trip is to be known, influential, wealthy, important, revered, ultimately enshrined in the supposed immortality of history. This trip is destructive to the individual, to the community, to the biosphere.

An alternate trip, certainly less destructive, is to scatter things differently, to *divest* yourself of materials, influences, casual friendships, relationships, dissipating activities, fame, wealth, and attention-stealing concerns, to learn to be less and less influential, less and less known, ultimately, of course, to *not be missed when you come to die.* To leave, perhaps, your total possessions, in the form of a bowl and a pair of chopsticks, to your grandchild.

But not to live or die alone.

To see you, to talk with you, to touch you, to make love with you. To feel and see and hear all heaven and earth with you, and through you, and in you. To make one whole life with you.

> I shall look for living crops from the birth, life,
> death, immortality, I plant so lovingly now.

WALT WHITMAN

NOTE 39. FREEDOM

> For it is not for what I have put into it that
> I have written this book
> Nor is it by reading it you will acquire it.
>
> **WALT WHITMAN**

Perfect freedom is the freedom to go everywhere, freedom to go into any field, any factory, any store, any street, any music, art, book, library, mind, into any world view, into any vagina. "Love" is being so completely into the other's knowledge, mind, background, emotions, experiences, loves, fears, delusions, *soul, that "into the vagina" follows automatically.* It was built to work that way.

And interestingly enough, the more you are into every other dimension of a woman (she is infinite, as is all nature, as we are too), the more you can get into the other dimensions, the deeper you can go into the vagina, and the better it feels. And the very best sex is when you can make love with the *whole woman simultaneously!* And she knows your body, mind, soul so well that she can make love with all of you, *simultaneously!* And a simultaneousness of those simultaneousnesses is the eternity that you seek and need.

Anything else is a "quick one."

The biologist in me (perhaps you could use that as a definition of what I think biology is) recognizes a difference in male and female psychologies—a very beautiful difference—that I can only think "into," into mind, into knowledge, space, vagina; whereas I think that women have the capacity to think "into" in the supremely womanly way, "into me;" women excel in receptiveness, personify receptiveness, are ultimate receptiveness. I only begrudge the fact that women cannot seem to understand that all we males can do is to intellectualize about that—we can just think about what it feels like to be receptive.

But of course, at the simultaneousness of simultaneousnesses, eternity of union, mutual perfect orgasm, you know that woman so well that when you think you know how she feels, you *feel* the way she feels. Otherwise you aren't in all the way. And the challenge, good friends, is to make it there, as far as she can go, which is a hell of a long way, because she is infinite in her capacity for pleasure. As the world is infinite in its capacity.

And as the highest perfect enlightenment is not to seek highest perfect enlightenment (not "not to have to seek," but "not to seek"), so is highest perfect simultaneousness of simultaneousnesses not to seek it (not "not to have to seek it," but "not to seek it"), *and yet to have it.*

NOTE 40. IT'S NOT SO FAR

Joe Jehl, Dave Hussell, and I in the cabin at Landing Lake, ten miles outside of Churchill, Manitoba; float plane in from Baker's Lake, Keewatin District, carrying an Eskimo with a bullet in his head. Car takes him and wife to the hospital; hot coffee in the shack with pilot and mechanic.

"Had some trouble up there, huh?"

"Yeh. Bunged the left pontoon setting down. What the hell could I do? Can't wait for you clowns to fly me some repairs, because the guy's lying there with a bullet in his head. So we hauled her up on this ice shelf, worked all night and patched it, and when it lightened up this morning we loaded him in there, I cranked her up, let her go, skidded down the ice about a hundred yards, off the edge, across the lake, and we lifted off. Didn't want to check to see if the damn thing leaked or not; not time to screw around."

"Well, you got him here anyway."

"Yeh. Don't know why these guys always doing something stupid like that during breakup, for chrissake."

So an Eskimo with a bullet in his head, having done something sufficiently spectacular and immediate was flown to the largest white settlement to receive the best surgical care that the military and civilian medicine of that frozen corner of Manitoba could provide. The wife looked dubious. Dozens of people die at Baker's Lake with a minimum of medical care. In the shacks on the edge of Churchill itself are Indians without any medical care at all. It's not so far.

In the tribal villages of central Africa, distances between huts or groups of huts are not measured lineally but in terms of how well the parties in question are getting along, how they may be related, how well they like each other. "Oh, it's not so far to such and such's" although it may be a good many miles; "We're good friends." "Too far to his place" means "I don't like that fellow." Short walk and long walk mean how inclined one is to go.

"I'm sorry, man, I can't make it to supper tonight."

"I'm sorry, woman, I can't pay any attention to you right now." Besides, a penis still bridges the gap.

"I'm sorry, beautiful day, I can't look at you right now."

"I'm sorry, soul, I've got to forget you now."

"I'm sorry, death, I cannot think about you."

Each year we spend more money on pet foods and supplies than we spend on nonmilitary foreign aid to feed our starving fellow men.

It's not so far.

NOTE 41. EVERYBODY HAS GOT TO TAKE ON THE JOB AT THE TOP

I do not make an exorbitant demand, surely.

D. THOREAU

The title paraphrases a quote from Jesse Watkins, whose trip into madness is outlined in R. D. Laing's *The Politics of Experience*. Jesse has entered a psychic state (called madness; Laing's thesis is that it is an exploration into the inner self, a search for greater awareness and for solutions to insoluble social stress) where he begins to feel like god when he has to take on the job at the top in order to understand for himself the unfathomable capacities of the human mind and the world—he has to take full responsibility for himself and his comprehension, full responsibility for his relationship to the entire world.

The type of perspective I think each human individual must acquire, the evolutionary and ecological perspective, would include a realization of the fact that your evolutionary origin means that the meaning of your existence is to be found in how thoroughly you can understand that existence, how completely you can realize your animal and human potential, how fully you can let go and explore the terrifying capabilities of your inner realities, how much you can realize and experience all the dimensions of conceivable inner and outer worlds. Those are the parameters of your evolved existence; to be unaware of some aspects of your potential is to be only partly alive. In the past we depended on absurdly weak excuses, gods, to understand things of which we were terrified but could imagine; we depended on some higher authority to prevent us from treading where all was uncertainty and fear. Now we realize that each man must be his own god, each man contains his own heaven and earth, each man must understand the dimensions of his worlds, each man is responsible for his own reality, his own realization, his own letting-go; ultimately perhaps his own madness.

We are not going to understand ourselves as enlightened human animals or behave in a manner that will begin to really let us see what we are and can be, that really puts us in an understanding and sympathetic and responsible relationship with the rest of the earth, until each man becomes responsible to his own task of understanding, becomes his own unfrightened god, *takes on the job at the top*, and is no longer afraid of himself or of anything else. There lies the only genuine sanity. There lies the only acceptable ecology.

NATURALISM

NOTE 42. WHAT ARE YOU LIVING FOR?

> I went to the woods because I wished to live deliberately, to front only the essential facts of life, and see if I could not learn what it had to teach, and not, when I came to die, discover that I had not lived.
>
> **H. D. THOREAU**

It seemed like it must have been a dream to George, or that he had just come out of a dream, or maybe he was dreaming now and had just started the dream. Anyhow, it seemed as if he was looking at life differently now, watching himself, as it were, wondering why he had not seen it this way before, wondering what he would think of himself tomorrow remembering himself standing there on the corner by the "Fast-Shop" wondering today if he was dreaming. Because it seemed like a dream now, or maybe this was real now, and before had been a dream. "Goddamn," he thought, "I'm not sure which of these is the dream, but they are different as hell, and I can't believe that they're both real."

George wasn't sure when it had come on. Maybe it hadn't come on all at once, maybe it was a gradual process, maybe it came on in little bits and pieces till one day it seemed that it was all there, that it had come all at once, that he had switched from dream to reality or from reality to dream. George wasn't sure. Maybe it was still going on, maybe by tomorrow or next week there would be more and then it would seem that a whole new dream or a whole new reality would emerge, all of a sudden, and be there, in him, on this corner, he, George, in a different, wider world. All that he was sure of was that somewhere along the way there had become fixed in his mind this question: "What am I living for?"

At first George didn't pay much attention. "What the hell," he said to himself, when he caught himself asking that question one morning while he was starting the car in the bitter cold and would rather have been in bed. "What the hell, everybody asks himself that once in a while. Nothing wrong with that." And knowing that the answer must lie safely available somewhere in that daily life of his, he let it go, forgot it, didn't let it worry him. But it wouldn't go away, kept nagging him, and more often now it would come back and he would find himself asking, "George, what *are* you living for?" "Goddamn," he'd say, "I haven't got time to worry about that kind of crap; got things to do!" But it stayed, and would not go away.

NOTE 43. WHAT KIND OF ENVIRONMENT DO I NEED TO BE ME?

> The whole civilized country is to some extent turned into a city, and I am that citizen who I pity.
>
> H. D. THOREAU

It is hard to escape the realization that the environment shapes the person. Is it possible that the ascendant *nurture* responsible for the development of personality can be in fundamental conflict with the *nature* of the organism undergoing the environmental modification, leading to a disastrous conflict between what we are meant to be as animals and humans and what we are made to be by the society within which we develop? Leading to frustration, meaninglessness, madness?

The contemporary urban society is utterly meaningless in the terms that give human environments and human activities meaning; the terms of what we are, what we have evolved to need and be, what sort of ecology we must have to be fully human. The selective processes that caused you to develop, body, mind, society, *the environments that called you into being*, were vastly different, naturally and socially, than the environment to which you are now exposed. You cannot help but be uncomfortable in an environment different from that to which you are adapted, different from that within which you evolved your meaning.

It is the contention of these essays that the organism is most interesting and healthiest when it is exposed to an environment fundamentally in harmony with the environments that, in the evolutionary process, were responsible for producing the bundle of adaptations that the individual represents. You need the environment you evolved in to be *you*.

And what we've got now, ain't it.

NOTE 44. LOOK WHERE I'VE GOT TO NOW!

> If a man does not keep pace with his companions,
> perhaps it is because he hears a different drummer.
> Let him step to the music which he hears, however
> measured or far away.
>
> H. D. THOREAU

One of the insidious and all-pervading diseases of this society, one of the basic environmental problems we have, if by this we mean undesirable aspects of our personal environments (which frequently happen, as in the present case, to generate problems in the external environment), is that we are educated to expect certain things of ourselves, to *become* something or other and to achieve a certain respectability, economic sufficiency, professional capability, power, etc. Our whole lives are patterned toward these achievements, and we are constantly placed in positions of invidious comparisons—not only comparing our success to that of others, our prestige and wealth to theirs, but also our success to our own unsatisfied longings, our own uncompromising ideas of how our personal worth should be expressed.

We will not solve the personal environmental problems or successfully stop the entire environmentally destructive sickness for material wealth and success until we ourselves learn, and until we teach our children, to enjoy life as it comes, to live for present satisfactions, to cease to dwell on future security and prestige, to take life for the immensely wonderful and mysterious process that it is. We must learn to live for the present, to minimize the planned and structured aspects of our activity, to now and again look around, and be able to say with wonder and pleasure, "Look where I've got to now."

A monstrous expression of our immense contemporary need for security in a world perceived as hostile is the intense interest in heavy insurance coverage; insurance for health, life, limb, car, fire, jewels, etc., etc., until there is no end to it. It is not only an expression of insecurity, it is also an expression of the wretchedness of the social organization and its unfortunate trend. The human personal environment is obviously not acceptable until there is, as there once was, no need for insurance policies. There was a time, not so very long ago, when if your house burned down, your neighbors built you another one, and if you died, they buried you and looked to the care of your family, and a man had no particular worry about unexpected misfortunes. The misfortunes still come in this heavily insured world, and between them you have the constant misfortune of apprehension about the sufficiency of your coverage. We shall not have solved our problems, we will not have an acceptable human environment, until the "insurance industry" is defunct.

Insure! Insure! Insure that you shall plod that wretched path for the rest of your "productive" life.

It's cold and dark and lonely outside the economic light.

How did we get along so far without it?

There is no insurance so cheap and as easily drawn upon as human insurance, as the respect and concern and affection of your fellow man.

Each time you buy an insurance policy you add to the deterioration of the social organism.

> The explosions are fearsome enough, but more so are the smoking slivers of stone that sing past your ear when the bolt crashes into a rimrock. Still more so are the splinters that fly when a bolt explodes a pine. I remember one gleaming white one, 15 feet long, that stabbed deep into the earth at my feet and stood there humming like a tuning fork.
>
> It must be a poor life that achieves freedom from fear.
>
> **ALDO LEOPOLD**

NOTE 45. "IT HAS TO BE DONE"

> I can feel
> all the shoes
> in a century of labourers
> being laced...
>
> **D. BLAZEK**

One of the conditions of a satisfactory personal environment is the ability to realize that there is nothing that "has to be done." To be able to sit on a bench in your yard watching the quiet evening come, with swallows "putting your house and yard into parentheses," at peace with yourself, satisfied, complete.

It seems clear that the way things are there is a great deal to be done. *The contemporary civilization must be stopped.*

But the dilemma is this: this feeling that it must be done, that we must get on with it, that we must try to change the world for the better (for ourselves, if we're selfish; for all mankind, if we get our kicks in altruism), that we must work for perfection, that we must have some all-consuming goal for this human endeavor, that the country or mankind has some purpose to achieve, this entire Protestant, indeed Western, ethic—*this impetus is the cause of our problems*; that's how we got into this bloody mess. And the question is this: do you fight fire with fire, can you use the destructive process to destroy the results of the destructive process? Can we generate a holy activism to stop an unholy activism?

It is the old story of environmentalists in a hurry, using the destructive system of air transportation to fly to national or international conferences on how to stop pollution, or inundation with airports, or sonic booms; the old story of housewives putting on their fur coats, jumping into their fancy cars to roar off to the supermarket to get some grapes (we trust nonboycotted), while insisting that the clerks put them in crumpled paper bags that they fish from their plushly lined pockets. Can you beat the system from within?

There is a real fear amongst concerned but not very perceptive "environmentalists" that a wave of apathy is sweeping over those young people who should be the up-and-coming soldiers of social change, an apathy of noninvolvement, an attitude of "it ain't my problem, lady." Well, the observation of apathy is in my opinion correct, but what is even more interesting is the fear.

The young people are perceptive as hell, and refuse to play these silly games at all, and they realize that one of the things that a person needs most of all is to be free of these anxieties. And the fear of the concerned people is to be expected, since they view the world and their relationships to it in terms of getting on with it, must do it now, etc., *just as does the bulk of mankind, the problem makers.* If they cannot out-work, out-endeavor, out-purpose the workers, the endeavorers, the purposers, they see ultimate disaster. That is disaster—to try to get your concerned work to progress faster than the destructive activities of the world; all you can do is to increase the pace, increase the odds that we will hasten to total unlivability. The opposition has the time, the men, and the money.

What is needed is a revolution in the direction of doing nothing, or very little. A revolution of underachievement to the point of mere existence, a total negation of the urge to do, control, remake, conquer, achieve, and a realization that this also means, it is the only way, to not participate in the destructive behavior either.

I would maintain that you have an obligation to try to make the world livable, to try to stop human destructiveness exactly in proportion to the destructive results of your participation in the ongoing process, and it is time to realize what some of the young realize almost instinctively, that the more you remove yourself from the process, the less you achieve, the less you work at anything, the fewer are the ramifications of your existence, the less disturbing the ripples of your being, the happier and less destructive you are going to be.

What you see as apathy, noninvolvement, underachievement, lack of motivation, the attitude that you in your monstrous insecurity fear; that is not a disaster but a hope, since it is the forefront of a revolution in consciousness and behavior that will ultimately destroy the rapacious orientation of western society.

We need more people to say: "I'm sorry, but I can't work on that, I'm busy thinking and living as simply and nondestructively as I can." Such an attitude, combined with a general enlightenment of what *is* destructive, is the only thing that can get us out of this mess.

NOTE 46. DON'T

> I suspect that a child plucks its first flower
> with an insight into its beauty and significance which
> the subsequent botanist never retains.
>
> **H. D. THOREAU**

It is important to be what you are at the moment. If you don't like that, don't be that, *now*!

The idea that we must do something because it is going to be necessary, some years hence, for us to have done (or learned) it, is a disease peculiar to this civilization. Where there is no security in what you are or who you love or who loves you, there is security only in having and getting and in girding your loins for a lifelong battle to achieve. It is a disease that totally permeates contemporary education, that fundamentally destroys the learning process—nobody learns anything anymore because they are genuinely interested in it but because they feel they must to get a degree, or be a decent human being, or some such thing.

Don't learn anything, don't be anything that you are not genuinely interested in learning or being. Learn because it is a pleasure to know something about yourself and the world about you, because it's a joy to see what the mind can do. I wish you luck at finding a teacher who learned for those reasons too.

NOTE 47. ALTERNATIVE REALITIES

YOU CAN GO AS FAR IN
AS YOU CAN GO OUT

DON CAUBLE

The world is as real as each person's perception of it. There are as many "real" worlds as there are perceiving individuals, and if it appears that many or most of these perceptions of reality coincide, that is only because we have discovered ways to minimize the social stress of conflicting realities; we have clamped the lid on aberrant noncooperative views, we have stifled the imaginative capabilities of man in search of the infinite dimensions of reality. This in itself is undeniably sad; it is sadder still to realize that the prevailing "reality," the reality of western civilization, the reality of growing up, going to school, getting a degree, getting a job, making money, "keeping" a wife, going to church, "getting yours," appears upon reflection to be fundamentally destructive, individually and environmentally. In other words it appears that measured against what we are, what we need, and what we could be, the prevalent "reality" is not real at all but an aberrancy, a diseased and distorted, incomplete perception of the biological reality of which we are a part.

The contemporary "real" world is *unreal*.

Possible realities are unlimited. This civilization is only one, and a crummy one at that. It is ironic, but perhaps to be expected, that this society should stand in such fear and trembling of those individuals who refuse to accept its perceptions of reality as real, of those who refuse to participate in that aberrant nonreality of social ignorance, of those genuinely interested in the limits of perception and imagination, of dope and philosophy. The only thing that this civilization does not fear is the thing that is most deadly, namely, itself and its horribly misinformed, miserably warped, totally inadequate system of reality.

The limits of reality are the limits of the human mind; it should be understood, of course, that this is part of the body, and the body is part of the earth's living fabric, and that the actions dreamed by the mind and performed by that body have got to be consistent with their maintenance on this earth. It is possible, in other words, for mind and body to destroy themselves; it is possible for the mind and its imagination, in its ignorance, to believe that it can escape the boundaries placed on it by the body and the ecosystem. We do that at our own immense peril.

This collection of essays is concerned with an enlightened animality and humanity, with finding an alternative reality, a social and philosophical framework within which the human individual can be healthy, happy, complete; an alternative that is concerned with a healthy and complete system for all living things, a beautiful, continuously functioning, rewarding biosphere.

NOTE 48. THE MOMENT AND FOREVER

> After the legalization era
> there was several hundred years
> of enlightenment.
>
> However that is where
> the records leave
> off.
>
> > showing total
> > preoccupation
> > with the
> > present.
>
> **RICHARD KRECH**

Progress is our most important pollutant. The idea that one is supposed to grow up in a constantly changing environment, in a social system rapidly changing itself and everything it touches—such an idea is a new one, but because we find ourselves caught up in it, we assume that this is the natural state of affairs, that this is a healthy social phenomenon. It is not. It is, of course, a profitable one, for intrinsic to our economic system is a need to grow; vast fortunes await those men who can get industries to grow, who can develop real estate and increase the insatiable demand for things. A stable nongrowth economy and a stable nongrowth population relationship with the surface of the earth are both fundamentally in conflict with our "free enterprise" economic system. An economist will say that the free enterprise system provides initiative. *That is true, and that is the problem*, for as long as there remains the initiative and opportunity to make a fast buck, the situation will remain unstable and can have no ultimate result except total destruction of the human environment. It is time to stop kidding ourselves; the present destructive progress cannot continue, and the present economic philosophy is not only *incapable* of changing it, it is *opposed* to changing it.

We need not only a new philosophy of life, we need also a new social and economic system. Free enterprise capitalism is a growth phenomenon; an eminent biologist has remarked "growth for the sake of growth is the philosophy of the cancer cell."

There was a time not so very long ago, when a man grew up, married, raised his kids, and died, in an environment that was continuously the same as that into which he was born. Today everything is in flux, and the resulting instability in our lives can be nothing but psychologically destructive. Why shouldn't you have an environment to live and work in where you would not have to worry about someone disturbing your peace and quiet, or someone building a freeway in your backyard or a root beer stand or girly show across the street, or dropping a bomb on you, or slowly, insidiously, poisoning you to death.

Physical and social stability are essential to psychic health. But our minds are plastic. I see no difficulty in envisioning an environment within which each person can have as much physical and social stability as he desires, but which allows the unlimited exercise of the human mind. We could have, in other words, a society that resists physical change (manipulation of the environment) while actively seeking new ideas, new philosophies, new fantasies. Our troubles result from the unlimited application of a single-minded economic philosophy.

It is not difficult to imagine an environment within which physical and social stability are combined with unlimited intellectual growth. Population growth and economic growth are killing us. Both, it is true, result from the growth of ideas. We are now in the process of generating ideas that will (must) restrict and eventually stabilize population and economic growth. It does not follow that the growth of ideas must be restricted. Their application, however, will be questions of immense social concern.

Each human individual should have the environment for which he acquired his animal and human characteristics. This demands a stable physical, economic, and social world, a nonhistoric, nonchanging environment, a continuous, happy, nonthreatened stoned existence. Eternity can be found here on earth. The moment of the individual existence could be serene and joyous; *then it would seem to last forever.*

NOTE 49. THERE IS NOTHING ELSE TO DO!

The mass of men lead lives of quiet desperation.
H. D. THOREAU

If I were asked why it is that I do not seem overwhelmingly depressed by the present situation or by the terrible future we are preparing for ourselves in our ignorance and greediness (sometimes, of course, *I am depressed*), I would answer that I have some trust in an enlightened human capacity, some hope that we will yet discover we should not waste our time getting things we do not need, that we should not waste our efforts building things we cannot use, that we need not do the majority of the worthless things this society demands of us, that we do not have to work for salvation, that we do not have to have so many things, *so many people*, to be happy and successful, and that we can and will, as a people, tell the present destructive culture to *stick it*. Then, interestingly enough, we will discover that we have the capacities and the time, the brains and the muscle power, to see to the maintenance of simpler life support systems, to see to the generation of personal environments of security and tranquility; we will discover that we have the time and inclination to live a beautiful animal existence.

In other words, it seems to me that if we abandon the foolishness, the destructiveness, and the insecurities of this acquisitive culture, if we abandon most of the ethics of the Western way of life, we will have the time and ability, the imagination and impetus, to find a satisfying and responsible ecology, to discover our potentialities and meaning in environments of ecological and evolutionary appropriateness. It turns out that *there is nothing else to do!*

NOTE 50. ORNITHOLOGY

> I learned today that my ornithology had done no service. The birds I heard, which fortunately did not come within the scope of my science, sung as freshly as if it had been the first morning of creation, and had for background to their song an untrodden wilderness, stretching through many a Carolina and Mexico of the soul.
>
> **H. D THOREAU**

Watching birds as a kid provided the basis for a genuine ecological sensibility.

It helped in the generation of a sense of wonder, beauty, and closeness to the earth and its suns and moons and spring smelling fresh of newly plowed soil and orange-rumped feigning killdeer on a spring day jumping on the ice candles of shore-thrown flood icecakes.

It helped by providing a background to biology, a feeling for the phenomena that biologists are trying to explain, so that when I finally became a biologist I declined to become a certified one, because those who are so busy at the "frontier" are so far removed from the phenomena they think they are explaining, *so far from the real frontier, which is to understand how good it feels to stand in the middle of the phenomenon and know that you are a part of it,* that I got bored with their lack of imagination and understanding. Not being able to ignore the beautiful Old Dame, life on earth in its fullness and diversity and me in it, I said to hell with them and left.

And it hindered me, since it made communication with people who cannot understand the earth, cannot feel its pulse or their own, who cannot stand on two inquiring and laughing feet to shout their joy of understanding the meaning of being alive and the meaning of spring to a school-hindered soul, made communication with house and city people difficult, and young maidens, too; ultimately making the oneness of a dual experience of earth existence difficult; although sometimes for considerable periods it has happened, and magnificently.

It hindered in the acquisition of a profession, a house, and security of the common sort.

If it has helped me to look to the earth for inexhaustible pleasure and inexhaustible interest and an inexhaustible sanity, it has also hindered me from becoming a civilized man.

To be quite frank, I like the hindrances.

To be quite frank, *I would try to hinder you.*
Or try to get you to let the earth and yourself hinder you.

> Wait, forget the Dean of Admissions who, if I came
> today in youth before him might not have permitted
> me to register, be wary of our dubious advice, step
> softly till you have tasted those springs of know-
> ledge that invite your thirst. Freshmen, sophomores,
> with the beautiful gift of youth upon you, do not be
> prematurely withered up by us. Are you uncertain about
> your destiny? Take heart, I, at fifty am still seeking
> my true calling. I was born a stranger. Perhaps some
> of you are strangers too.
>
> **LOREN EISELEY**

NOTE 51. DIVERSITY

> I seek acquaintance with Nature—to know her moods
> and manners. Primitive Nature is the most interesting
> to me. I take infinite pains to know all the phenomena
> of the spring, for instance, thinking that I have here
> the entire poem, and then, to my chagrin, I hear that it
> is but an imperfect copy that I possess and have read,
> that my ancestors have torn out many of the first leaves
> and grandest passages, and multilated it in many places.
> I should not like to think that some demigod had come
> before me and picked out some of the best of the stars,
> I wish to know an entire heaven and an entire earth.
> All the great trees and beasts, fishes and fowl are gone.
> The streams perchance are somewhat shrunk.
>
> **H. D. THOREAU**

One reason for maintaining organic diversity, for maintaining some semblance of a natural environment, is to prevent the shift to human activities, distractions, and recreations that we must otherwise make to keep from being bored to death. This shift is self-accelerating and self-reinforcing for two reasons: first because the diversions are so unsatisfactory that they lead to even greater boredom, and second because these attempts to provide ourselves with diversions from real life in the natural world result inevitably in less diversity. So the natural environment and our activities within it become poorer and poorer, while our human world becomes more destructive as well as unfulfilling and meaningless. The question is whether or not we can stop it, whether or not we can reverse it by finding more and more excitement in more and more natural diversity, until we are living a totally satisfied, totally aware, totally stoned natural life.

Our animality and our humanity demand that we exercise them. *They will not be satisfied until they are exercised in those situations in which they were evolved.*

A sterile environment can only product sterile souls. Diversions and diseases such as schizophrenia are attempts to find sanity and health in unacceptably deteriorated environments. As long as we need madness and diversions as retreats have we solved our environmental problems?

NOTE 52. SAVORING

> What tribe is this so sunk in itself that it dreams
> in a night gone crazy.
>
> **KEN KESEY**

And why do I write these exhortations? Why do I bother to try to teach you that your civilization is destroying you and itself? Why do I try to suggest that a rational world view, based on evolutionary biology and ecological principles, must replace our irrational, superstitious, and materialistic culture? Why do I rub my face in this dung? Why don't I, too, say to hell with it? Well, it's because I want to dream a genuinely pleasant human-animal dream, and it's hard to dream in a night gone crazy.

I would rather sit in the late afternoon sun on the bench in my yard, listening to the evening songs of birds, to ruffed grouse drumming in the distance, to insects swarming around the trees and flowers, to the comforting low calling of my geese, to the lowing of the cows in the pasture, maybe to a barking fox or two, and with luck, to a whippoorwill when the beautiful moist twilight comes. I would like to see barn swallows enclosing my house and yard in the ellipses of thier flight. I would like to be able, each spring, to see the hooded mergansers and golden-eyed ducks, and geese and dozens of kinds of shorebirds, and thrushes and warblers and hundreds of other kinds of birds wing their ways northward. And I would like to see another forty bloomings of spring pasqueflowers and bird foot violets, forty more summers of prairie fullnesses, forty more autumns of nut-ploppings.

I would like to continue to have the time and place to grow my own delicious springtime radishes, plump summer cabbages and tomatoes, sweet corn that you cannot believe, with no end to it; to have the time and inclination to plant every flowering shrub with a fine odor that will grow here, a spring and summer and autumn of fragrant shrubs. And I would watch my bees and look forward to the summer when could I plant a whole field of buckwheat for buck wheat honey (and the pancakes!).

And I want always to be able to find an undisturbed boreal forest to snowshoe through, a calling loon summer wilderness free from the sound of motors, and maybe someday see the tundra and the tropical rainforest again.

And I need not only the peace and quiet and beauty and health to be found in living on the land, with the land; I need also the human community of a noncompetitive nonmaterialistic social group, a cooperative human endeavor, an affectionate and personal human society. Everyman needs this.

And I would have the time and inclination to encompass as much of the human experience as possible: ideas, books, songs, histories.

And I would have a child and have him or her have a future, as I would every man and woman and their children have the same.

And I (and you) cannot have these things if I or you or the rest of us keep going on the way we are.

NOTE 53. TRIPS

> Cultivate poverty like a garden herb, like sage.
> Do not trouble yourself much to get new things,
> whether clothes or friends.
>
> **H. D. THOREAU**

Alcohol, tobacco, dope, sex, the nicely rounded ass, the pert and pointy breasts, the record players, fancy clothes, sheer nylon bikini panties, big fancy houses, landscaped yards, straight white teeth, 240 cubes under a moulded purple hood, radials, the new rug, antiques, Ph.D. degrees, movies, promotions, body adornments, refrigerators, freezers, dishwashers, Vick's Vaporub, orchid-flavored lip balm, eye shadow, Danish modern, silver-plated, queen size, thermostatic, automatic, flavored and colored, pre-cooked, predetermined, strong, powerful, busy, agri-busy, agribusiness, development, progress, empire, snowmobiles, one night stands, football games, crackerboxes, cupid's quiver, Coca Cola, cordless massagers, contests, affection, thrust, faster, new, mine, me, good, new, better, mine, me, me, me, me, me, aaaaaaaah!

Everybody is tripping.

You can trip on almost anything. You can trip the cheap trip of the American dream (it's a bummer). You can trip on Jesus (intellectually, emotionally, and, I maintain, environmentally, that's a disaster). You can trip on nature. You can trip on your inner self. You can trip on beauty. You can trip on another person. You can trip on the bright lights of the city. You can trip on Christmas cards. Have you ever wondered at the silliness that, once each year, for no particular reason, results in a massive social attempt to destroy the Post Office? The men who fashioned the postal service, probably unwittingly, set it up so that costs and volume adjusted themselves to allow the fullest tripping possible by written means. Have you ever known someone who didn't write because they couldn't afford the postage?

Telephone in every bedroom!

It is no particular wonder, really. During the evolution of that massive brain, in the primitive tribe with all its magnificent complexity of interactions, in an environment of unlimited diversity, in the excitement of the chase, in the fear and trembling before an unfathomable spirit-filled world, in the superimposition of minding and aesthetic appreciation upon the deliciousness of sex, the capacity for ecstasy became part of our humanity. In an environment far less interesting, far less satisfying, and many times more stressful, to keep from going mad *you have got to trip on something.*

A principle of an evolutionary ethic is that we must meet the needs of our total animality and humanity. If we are made to trip, we must let "it all hang out." A principle of an ecological ethic is that *we must engage in nondestructive trips.* The ecological perspective provides us with the realization that this is one interacting world; behavior that is destructive to other humans or to the surface of the earth is ultimately destructive to the person tripping. A destructive trip is self-defeating. A self-defeating trip is ultimately a bummer. We will soon be dead of our own too much (Leopold).

There are all kinds of beautiful, nondestructive trips. Find some.

NOTE 54. DOPE

> I answer that I cannot answer, you must find
> out for yourself.
>
> **WALT WHITMAN**

The availability and widespread use of alcohol and marijuana, necessitated by otherwise unsatisfactory personal environments, allows us to treat the symptoms of our social malaise and therefore keeps us from recognizing and curing the disease. Without dope we would *have to* change the society; with it, the present abomination is sometimes almost tolerable. "I gotta have a drink."

Or, if it is innately animal and human to need to trip on something, then dope can be used to provide nondestructive trips and should perhaps be substituted for the vainglorious and environmentally destructive trips of this civilization. "The joint is my trip, not the presidency of General Motors."

Dope can take you back to your animality; it can show you some beautiful aspects of your primal nature. It can teach you that your mind and thoughts and deepest feelings are the productions of the chemical operations of your brain, because it is that brain that is affected by the chemicals you smoke or drink. It *does* provide a chemical soul; or, more properly, it affirms what a biologist believes, that the soul *is* the chemical operations of the brain and its capacity for memory and experience.

Dope teaches you how beautiful the world can be. The realization of that inexpressible beauty, together with the experience of the wretchedness and meaninglessness of the contemporary culture, necessarily results in disaffection. Who cares to participate in a society designed to keep you unhappy? The immense outcry about marijuana use must be understood in the context of the tremendous threat it poses for the existing order; I would maintain that its use and the resulting perception of how beautiful the world *could* be is the single most important challenge to our economic and social systems. The conflict is between those who would maintain the present disaster and those who believe that life is meant to be healthy, happy, and simple.

If every "head" returns to dope, or to a situation that provides him with the peaceful ecstatic feeling that dope produces, *as he must* (and when he is stoned he realizes this), then we will eventually end up with a "stoned" culture, a culture of ecstatic existences; *and that is the beautiful aspect of the drug culture.*

Childhood is a stoned time, and old age is, too. Why can't the middle years also be beautiful, happy, stoned, careless? *We have not solved the environment problems until they are.*

Dope allows you to forget that massive oppressiveness of reacting mind, of constant intellectual searching, and lets you see the simple "is-ness" of things and of yourself.

NOTE 55. GO!

> For we are bound where mariner has not yet dared to go,
> And we will risk the ship, ourselves and all.
>
> **WALT WHITMAN**

It is obvious that we are programmed to *go!* There are, I suppose, societies that are relaxed and nonachieving, but I am afraid that our emphasis on progress, our insatiable thirst for new and bigger and better and more, is not just a cultural phenomenon. We might be more successful if we looked to our psychological makeup for its origin, if we came to realize that our present difficulties stem not only from some misapplication of technological proficiency but maybe even from the depths of our animal and human personalities, from capacities unleashed in unfortunate directions, needs now unfulfilled and prodding us to desperate searchings.

Somewhere in the evolution of our sociality we developed a gregariousness, a tolerance, indeed a necessity, for close contact with our fellows, a need for cooperative activity, a need now expressed in the teeming masses of cities, a need, perhaps, that will culminate in the utter blindness of massive overpopulation, of total, all pervasive interaction (and madness). Somewhere in the evolution of our sociality we developed or refined an inordinate curiosity that is expressed today, interestingly enough, by an insatiable hunger for news of the human world and a total lack of interest in our nonhuman surroundings. Somewhere in the evolution of our sociality we acquired a frenzied restlessness; adrift and alone with a developing culture that began to mediate between ourselves and an unfathomable environment, we began to feel our oats, began to dream, began to climb a ladder to the stars.

We cannot stop the madness that is our humanity.

But we have got to find nondestructive ways to go, to let it all hang out, to trip on life, to be utterly irresponsible and revel in our capacities, in the unimaginable freedom that is ours. How stupid to bung up the process with ignorant and inadequate trips: that is, too many babies and a rampant materialism.

Sometimes I want to say: "I could trip better, you could trip better, all men and animals and plants could trip better, if you would just not screw it up!" This the meaning of pre-emptive resource use.

But that approach is insufficient; nobody likes negative reinforcement, nobody likes to be told "No!" We will win this monstrous fight, horribly outdistanced as we are by our technological joys and plastic cities *and carefully cultured human stupidity*, we will win it only if we can get interested in the natural world and natural phenomena and ourselves as living organisms and say GO! and enjoy yourself and the beautiful earth! GO! GO! GO! And do not be polite and unoffending, do not be programmed, do not waste your time at uselessness, do not settle for cheap pleasures, *learn to live what you are.*

And this trip toward naturalness, this trip toward living in and appreciating the biosphere, this trip toward ecologically and evoutionarily responsible behavior, this trip toward simplicity, exuberance, freedom, wholeness, oneness with the earth, this trip, unlike some others, *is good for you.*

If somebody else's trip destroys your high, you had better get him high on your trip, or he will not stop. When your high is genuinely his high, he will have to stop what he is doing to keep from hurting himself.

An important aspect of an ecological ethic is to explain your high, to help others to see and feel and understand the world and themselves.

You have to see it, feel it, smell it, touch it, understand it, savor it, and then *let GO!*

<blockquote>
To be lost if it must be so!

WALT WHITMAN
</blockquote>

NOTE 56. CHOCK FULL OF STUFF
TO GET YOU OFF

To be conscious of my body, so satisfied, so large!
To be this incredible God that I am!

WALT WHITMAN

I remember the countless thousands of times, as a kid, when I went into the bathroom to pee and found one of my father's cigarette butts in the toilet. I used to chase them around the bowl with my stream of urine. But I remember now, particularly, the dark yellow stain trailing down the edge of the bowl from the floating butt. Nicotine and tars. A marijuana roach in the toilet does the same thing. They are both chock full of stuff to get you off.

So the Surgeon General's report demonstrates conclusively to any individual who can understand the methods and conclusions of scientific inquiry that smoking tobacco will kill you. And why don't people quit? Is it because they have an attitude that "it ain't gonna happen to me," or, being good Christians and expecting eternal life elsewhere, they don't give a damn? *Or is it that they have to have a high*?

The old man's dead now, of lots of things, but the tobacco contributed its share to the destruction.

Twenty highs to the pack. "Plus Tax."

NOTE 57. I COULD HAVE STAYED AT HOME WHERE THERE ISN'T ANY WATER

> The old naturalists were so sensitive and sympathetic to nature that they could be surprised by the ordinary events of life.
>
> **H. D. THOREAU**

Now and again I go to the wilderness in a canoe. I'm particular about how I go, however, because I don't want to take the technological world with me. I go in a cedar canoe, beautifully made of wood and canvas and varnish and human hands and love. Such a canoe, in materials and workmanship, is a direct and legitimate offspring of the finest of all North American aboriginal craft, the birchbark. It pushes its quiet way across a wilderness lake in a manner that can only be achieved when men and materials are natural to the context, and behave softly, quietly, organically at home.

An aluminum canoe is something else again. The bowels of the earth are ripped open, and bauxite is removed. (Will you dig it or spend your alloted portion of living experience building the huge machines that do this monstrous work?) The bauxite is transported to a refinery. (Will you transport it; will you refine it?) Huge amounts of electricity are used to reduce it to metallic aluminum. (Will you dig the coal or drill for the oil; will you breathe the fumes of its generation; will you furnish the steel and the swaths for its transmission; will you lead the accompanying lives?) The molded sheets are sent to an aircraft factory, slipped into huge presses:

Wham! (One canoe half.) (Will you stand at the press?)

Wham!   (Another canoe half.) (WILL YOU STAND AT THE PRESS?)

Rat, tat, tat, tat, tat, tat, tat, tat, tat, tat, tat, tat, tat, tat, tat, tat, tat, tat, tat, tat, tat, tat, tat, tat, tat, the rivets are driven home (ah, the quiet of a wilderness lake.)

Wham!

Wham!

Rat, tat, tat, tat, tat, tat, tat, tat, tat, tat...

To canoe in an aluminum canoe is like making love with a dildo.

I would push my entrance into the calling stillness of the mysterious wilderness in a craft composed of wood and canvas and men's hands and hearts and imagination, an organic man-vessel.

True, the aluminum canoe does not leak. But if I merely wanted to keep my rear end dry, I'd have stayed at home in my backyard where there isn't any water.

NOTE 58. WHAT CAN YOU AFFORD?

> I have these wonderful things
> to give to you if you will
> promise
> to sing me a song about how
> everything was before
> this world stuck its fat ass
> in the window and shit
> all over the table.
>
> **D. R. WAGNER**

We have to afford what is cheapest for us and therefore most costly and destructive in an environmental sense, because we have to afford so many other things or think we do. And there lies the cruel irony of manufactured "wants," for they give us crap too, which we don't want at all. If you want the nice things, you ought to take the crap, not make someone else live in it. I, for one, do not want to live in it, friends; I guess I have a low crap quotient, or maybe I just have a better crap detector. Interestingly enough, I find that if I learn not to want your fancy things, then I can quit working in town and live and work where there is hardly any crap at all except cow manure, which I can use. Except, of course, that there are now so many of you that the crap is flying all too thick out here, too.

These, the fruits of wanting, "affording" things.

You can't afford them; they will kill you! If not your body (they'll get that too), then your soul.

No, I guess they already have.

NOTE 59. I NEED MORE

> That man is richest whose pleasures are the cheapest.
>
> **H. D. THOREAU**

We surround ourselves with all kinds of junk we do not need, all kinds of things for which we trade our lives, using them to buy more things, things that steal our attention from other living creatures, humans too, lovers maybe; things that separate us from what we are and from where we are, no end of things in which to hide our fears, in which to lose ourselves, things to ease the burden of uselessness, things to conceal the ignorance.

So many things.

What can you do without?

"I need all these things to live!"

"I need all these things to be *me*!"

You don't need any of that crap. Or you would not need it in a world of more primitive social groupings, a world that provided not only identification and personality and birth-to-death security and affection but also a great deal of interesting things to attract the attention, people, animals, *life*.

Houses full of things ("I need more!").

Houses full of crap ("I need more!").

How much do you need? (Friskies *and* milk for the cats?) (Sure, there is lots of milk. We'll add another cow, if necessary, make more hayfield if we must, use more of the earth if we have to, to *make life more livable* for these cats.) (And thus for ourselves.)

(How livable?)

(How *many* cats?)

How livable, and for how many of us, too?

NOTE 60. DO YOU LOCK YOUR DOOR?

> A lady once offered me a mat, but as I had no room to spare within the house, nor time to spare within or without to shake it, I declined it, preferring to wipe my feet on the sod before my door. It is best to avoid the beginnings of evil.
>
> H. D. THOREAU

Stayed at my friend Skinny John's place one weekend to feed the animals, look after the heat, etc., while they were gone off. Had to go into town Saturday myself, so I asked if I should lock the house while I was gone. He turned to his wife and asked, "Did the real estate man give us a key when we moved in here?" "I don't know," she said. "Do we have a key?" "I don't think so." They had lived there for eight years! In all the time they had lived there they had not locked the house, could not lock the house, *had no key.*

And so you learn, slowly and to your pleased surprise, that even close to populous cities there are communities wherein a man's place is respected by all his fellows, where there is nothing to fear, no need to "lock up," day or night. Do you think that you evolved in a community where you had constantly to fear your fellow man? Do you think that the locks will keep him from stealing your psyche? If you have a lock, he has already destroyed you.

No man should own anything that another man would want to steal. Which is another way of saying that every man should have what he needs and *learn not to want more.* Until such a time as no man needs lock up anything, the environment is degraded accordingly. Pollution control as presently defined does not even scratch the surface of the problem. Possession and theft are *both* pollutants of the human spirit.

NOTE 61. THE PLEASURES OF POVERTY

> Now I see the secret of the making of the best persons,
> It is to grow in the open air and to eat and sleep with the earth.
>
> **WALT WHITMAN**

There are two reasons for not acquiring material possessions: the reason of ecological responsibility, and that of personal liberty and development.

It should be clear by now that there is a limited number of resources on the surface of this earth, and therefore one should have some preliminary realization of limited goods per individual, there being such a godawful large number of us. A more sophisticated analysis leads to the realization that there have been some horrendous consequences of our attempts to rectify the shortage situation by our fancy technology, governmental organization, etc. In our attempts to make the acquisition of enough to eat easier, we have destroyed most of the soils of the earth; we have reproduced like hell, with the result that most of us are undernourished, and Western man, even with an abundance of food, is malnourished, rife with degenerative diseases, and beginning to show a drop in life expectancy. In our attempts to manufacture the good life for everybody we have deprived ourselves of fundamentally important aspects of living, of clean air and clean water, open natural spaces, quiet in the night, darkness, serenity, individual purpose, and opportunity to perceive beauty. A fundamental ecological principle of human activity on this earth is this: *we can make an unlimited amount of anything only with the inevitable, irretrievable loss of something else that we must have to be healthy and happy.* The massive human activity and organization needed to carry on our complex civilization produces, by its very nature, an environment antagonistic to the good life. The prevailing theme is "we can do anything," and it sometimes seems that that is so; it is important, however, to accept the obvious fact that anything we do has massive environmental implications, almost all of them bad.

The American way of life is criminal. Those who think that we, a small proportion of living persons, a small segment of the humans who now are experiencing or will experience this earth, have the right to destroy most of the important features of the earth environment so that we can have a big car, a fancy house, a new refrigerator, a fancy stove, a hi-fi, a tape recorder, a closet full of clothes, unlimited energy and transportation, are all of them criminals. We are six percent of the population of the earth, and we annually use something like sixty percent of all the minerals and energy that are extracted from its surface. The political radicals call this "American imperialism," rightly so, I think; I call it massive ecological criminality. From the point of view of ecological ethics, it is a criminal act to own and drive a big automobile, to live in excess of your needs. Since there are so many of us engaged in such excesses of demands, reveling in such luxury and disgusting opulence, it is hard to accept the realization that we do so by denying other persons what they need for survival and by bringing about a deterioration of the entire earth ecosystem. *The American way of life is a criminal act.* Within the framework of a realization of the limits of the earth, the immense numbers of us, the needs of each of us to become healthy and happy humans, and the need to continue our life support systems indefinitely into the future; within this framework any man who owns more than the necessities of life is an *ecological criminal.*

There are some pleasures to be found in self-imposed poverty. There is one immediate and important reward that follows from learning to do without the unnecessary trappings of this society: *you don't have to work so hard.* Thoreau found that he could live on six weeks labor per year. Certainly there is no excuse for working at "bread-labor" for more than six months of the year. The rest of your time is yours to live life, to see and feel and enjoy.

There are some even greater pleasures. In this society the very nature of reality is defined by the economic sphere; this is a very curious invention of Western society. Economic considerations impinge so fundamentally on human existence in this culture that they literally encompass and become the meaning of human lives. To whatever degree you escape the economic world, as an earner or spender, to that degree can you search out and adopt alternative realities, alternative systems of philosophy, of thought and action, perhaps even of fantasy. It should be unnecessary to suggest that any number of these unlimited alternatives are infinitely more interesting and satisfying, positively less destructive to one's person, to other persons, or to the environment, than is the prevailing rat race for economic superiority and oversufficiency.

Unless you wish to become dependent on another person, or remain so if you now are (that is its own trip, of course), you have got to remain sane and healthy enough to make a living and care for yourself. If you are satisfied with a simple living (or if you are very rich, I suppose; I must condemn this because I am convinced that it is destructive to the individual, to say nothing of the individuals, indeed generations, that have been exploited to provide that richness, and also because it is destructive by virtue of encouraging the almost universal doctrine that the only way to security *is to work hard and get a lot,* so everyone, in his ignorance, does so), if you are satisfied with a simple living, you can be damn near as mad as you wish. Which is to say you can participate in all sorts of realities outside the economic sphere.

There are further, more fundamental pleasures. If you live deliberately poor, you are likely to live close to the earth and continuously experience the animal joys of securing food, finding rest, developing your skills in growing food, obtaining inexpensive clothing. You may learn to depend on your fellow man, securing in the process the most important of all human achievements, *health, dignity, trust and affection*. There is wisdom in insecurity (Alan Watts). And you will have all the rest of your time, all the rest of your life, to see the daylight come and the evening darkness settle upon the earth, to play with your children, to fondle your lover, to learn and to see and hear and smell and appreciate. To live a full animal and human existence.

All you have to do is to learn to say to hell with it! And take the consequences. To learn to satisfy your economic needs simply, without much bother or wasted time, without selling your principles or your soul. It is possible to learn to live in poverty with dignity. In fact, I maintain that it is impossible to live with dignity in any other way.

Make the break to a happy, healthy, environmentally responsible life.

Get thee povertied.

NOTE 62. PARAMETERS OF AN ACCEPTABLE HUMAN ECOLOGY

> Now I am terrified at the Earth, it is that calm and patient...
> **WALT WHITMAN**

An acceptable human ecology must do two things: it must provide each living human individual with a satisfying environment, and it must develop an harmonious equilibrium relationship to the rest of the living world, to the surface of this earth. It must, in short, provide a reality alternative to that which we now perceive and accept. What would be some of the conditions of such a human ecology, of such a reality?

It would provide each individual with the physical, chemical, and biological necessities of life, *but only the* necessities, thereby assuring that all men, now and in the future, will have them. It would accept the fact that there are too many of us, that each of us wants too much, and that the solution for that aspect of an harmonious ecology is obvious—we must see to it that there are fewer of us, and that each of us wants, and gets, less. It would be an ecology of *learning to share* and *doing without*.

It would provide each individual with the psychological, social, emotional, intellectual, and aesthetic necessities of life, totally fulfilling his needs in these respects, developing his unlimited capacities for empathetic animalness and humanness. It would provide more than the "necessities" in these respects, because it could do so to an infinite degree (I should hope by now to have convinced you that we have infinite necessities in these respects), since none of these needs are limited by considerations of energy, manipulation, or natural resource depletion. The most pernicious depletion of environment is that which follows from unlimited population and material wants. Not only are these wants inevitably frustrated, this economic and material want-fulfilling also destroys the capacity of the natural and social environments to furnish the psychological, intellectual, emotional, and aesthetic necessities; it depletes the environment of these qualities, steals what is otherwise an unlimited, nondepletable resource.

An acceptable human ecology would be an ecology of learning to share an inexhaustible content.

It would treat all the substances of the earth as sacred, precious, to be carefully used and recycled, *if used at all*. It would release no substance to the environment that could not naturally and harmlessly be reincorporated into the living world.

It would be a human ecology without a pressured economic sphere.

It would be a society wherein each individual lived in direct contact with his physiological and psychological life support systems, close to the earth, in contact with plants, animals, clouds, seasons, rains, snows, and the source of his existence and meaning. It would be an ecology of an active hardworking healthy individual with a worldwide intellectual life.

It would provide each individual with the infinite pleasures of a noncompetitive, nonsuperficial, secure social grouping.

It would provide each person with an opportunity to develop and live a biologically sound philosophy of life, to revel in his animism and forsake his fears and superstitions; it would provide the opportunity for an infinitely pleasurable love affair with life and with the earth.

We must stop doing that which denies us such an ecology. And that means that we must stop the present philosophical and economic insanity, we must stop being ignorant, superstitious, and greedy.

These essays do not constitute a handbook of steps for social change. They attempt, instead, to provide an ecological and evolutionary perspective, a philosophical basis for the necessary social enlightenment. How to get from the philosophical position to the responsible social state is, obviously, an extremely difficult question. I would not be so foolish as to prescribe the necessary behavior for anyone but myself. But it seems clear that each of us has the responsibility to change himself, to abandon what he thinks and does if these are inconsistent with an ecology acceptable to all men and to the earth.

> If I have got to drag my trap, I will take care that it be a light one and do not nip me in a vital part. But perhaps it would be wisest never to put one's paw in it.
>
> **H. D. THOREAU**

NOTE 63. THE FUNDAMENTAL RULE

An elderly dame, too, dwells in my neighborhood,
invisible to most persons, in whose odorous herb
garden I love to stroll sometimes, gathering simples
and listening to her fables; for she has a genius of
unequalled fertility, and her memory runs back farther
than mythology, and she can tell me the original of
every fable, and on what fact every one is founded,
for the incidents occurred when she was young. A
ruddy and lusty old dame, who delights in all
weathers and seasons, and is likely to outlive all
her children yet.

H. D. THOREAU

Live as simply and as naturally and as close to the earth as possible, inhibiting only two aspects of your unlimited self: your capacity to reproduce and your desire for material things.

NOTE 64. DEATH AND RECYCLING

> All goes onward and outward, nothing collapses,
> And to die is different from what anyone supposed,
> and luckier.
>
> **WALT WHITMAN**

Sometimes I think that the ultimate trip is to go mad, to see what that's like. But I know that there is an even better trip, and I'm afraid that thoroughgoing madness might deprive me of the opportunity to savor it. *That trip is watching yourself die.* It seems clear to me that to die instantaneously is to be cheated of a most valuable experience. To have the pleasure of watching yourself cease to exist seems to be the ultimate trip, the culmination of a lifetime of experiencing. It should be the most exciting pleasure of your life; it is unquestionably the last.

(Thus it is that I cannot stand civilized society; I tremble at the thought of instant death at the wheels of a garbage truck.)

The most obscene manifestation of our refusal to accept ourselves as animals, as part of the living productions of this good earth, is our nearly universal attempt to keep ourselves from rotting back to the soil that produced us. To this end you will be saturated with embalming fluid, sealed in a steel box, that placed in concrete, and the whole buried too deep for the roots of grass to suck you once more into the sunlight.

"I hate you earth, so much; I fear you body, so terribly; I tremble so mightily at the prospect of dissolution; I scream and hide from that awful realization; I must be preserved; I want to *be*, I want to *be*, I want to *be*.

Well, *you* will not be, just as *you* were once not, but your atoms will go on, just as the atoms that have already been through you, and *they* will be, and whatever they are part of will realize them in its own quaint way, and differently from the way you now realize them, and they you.

Sometimes, hurrying to town, I am reminded of the ridiculousness of my haste when I pass an outcrop of Ordovician sandstones. For five hundred million years those rocks have laid there; where they are exposed or as they become parent material under developing soils, they are slowly decaying and releasing their atoms. The cabbages in my garden are Ordovician in their mineral origin. I am made of these ancient sandstones. For five hundred million years they have rested. Should the fierce intensity of the oxygen that I have combined with these ancient minerals so agitate me as to make me think that I have someplace to go in a hurry, something to do, and I've got to get on with it? And my atoms, and yours, will eventually rest even longer in the primeval seas and ancient strata of unbelievably distant future eons. Why so hot? Why the anxiousness and the fear?

The test of the completeness of an ecological and evolutionary ethic is whether or not you can accept yourself as a collection of earth's atoms, assembled by the marvelously complex, wonderfully interesting, natural processes of this planet; whether you can achieve the peace of understanding your existence and your ultimate disintegration, the peace of a full and beautiful understanding of the fact that the forms are seen and forgotten, but the process, the phenomenon, passes on.

There is no meaning to fear; there is no death but that of the self.

>   I bequeath myself to the dirt to grow from the grass
>   I love,
>   If you want me again, look for me under your bootsoles.
>
> **WALT WHITMAN**

NOTE 65. A NATURALISTIC ETHIC

> I only am he who places over you no master, owner, better, God, beyond what waits intrinsically in yourself.
>
> **WALT WHITMAN**

You are an animal, part of a natural process, part of the evolution or convolution of the universe, part of the living fabric of this earth, part of a process whose origin was inevitable in the bonding potentialities of this planet's atoms.

You evolved here, body, mind, soul.

You are merely one specimen of one of the millions of living forms, no more important, no more beautiful, no more meaningful than the rest. You are the result of the historic pathways that your gene pool has followed in its complex and intricate journey from simple living stuff to yourself, just as every other organism is the result of the paths of its gene pool. The only meaning that you have is found in that pathway, just as the only meaning of all other living forms is to be found in their pathways.

You can be a healthy animal only if you are exposed to the environment to which you are adapted, within which you evolved, within which you have your meaning, within which your capabilities can be expressed. You can be a healthy *human* only in an environment in which your humanness is meaningful, an environment within which your humanness was called forth, evolved, only within satisfying and secure social relationships.

You must have these environments, and you must not destroy them by your behavior within them.

You come from the natural environment, and you will go back to it; that is, parts of the environment become you (as when you eat, breathe, see, feel, or think about it), and when you are done with them, parts of you become environment (as when you urinate, etc.). You are a node; when the *node* no longer is, *you* will no longer be. You are rocks, water, and air in a *thinking* combination.

You are *universe aware of itself.* Each person is a unique way of universe aware of itself. *Each is universe.*

The natural world is infinitely complex, and neither you nor any other man will ever understand it sufficiently to control or improve upon its processes.

You are your only limit to the beauty, pleasures, knowledge, wisdom, fascination, serenity, and love that you can receive from this earth and all its processes, forms, minds. You, that is, and any behavior of yours or of another man that destroys some of earth's beauty, pleasures, knowledge, wisdom, fascination, serenity, love.

Your capacity for empathy with the earth and all its organisms, its rains and clouds and storms and seas and mountains, is limited only at your end of the relationship. The earth has limitless empathy, she will take you and me back one day, no matter how repulsive and destructive we become, but her capacity to provide for us now is severely damaged when we beat on her.

If you fail to develop that empathy with the earth and its processes you deprive yourself; you live a life that is less than fully alive, and your ignorance will eventually destroy you (and yours) (and me) (and mine).

Watch and appreciate your animality and your humanity.

Realize that meaning, in the present context, is to be found in providing yourself with an environment suitable to the complex bundle of adaptations that you are and in participation in a social effort to bring your species into a sustaining equilibrium relationship with the earth.

Yourself is one-half of the reproductive and psychic Self.

You are a mind-beast, but you must not let your minding interfere with a complete expression of all the other aspects of your humanity and animality.

You should realize that meaning is to be found in letting yourself go and in how fully you can explore the terrifying capabilities of your inner reality; genuine sanity and an acceptable human ecology will only exist when each man takes on the responsibility of fully understanding himself.

You need to "trip." But because you and all other men must continue to have the opportunity to trip, you must find some nondestructive trips, lest yours and theirs become bummers for you or them or men and women yet to come, yet to have those glorious possibilities of existence.

You must seek a whole heaven on a whole earth. You must realize that in order to obtain it, you and all men must exercise an ecological responsibility, must inhibit two of your unlimited abilities, the ability to reproduce and the desire for material things.

You must lead a simple, nonacquisitive life, close to the earth, so that you will minimize your impact upon it.

You must realize that there is no death but that of the self, that the ecological and the evolutionary processes go on.

We are all lost, friends, but some of us are no longer afraid.

I promised myself that I would be finished with these essays when the redwinged blackbirds came back, and yesterday they returned, and it is time to quit this nonsense.

Can you still believe that environmental problems are encompassed by cleaning rivers and reducing automobile emissions? Can you still be so naive as to think that this is anything other than a fundamental question of what we are and how we got here and what we might become? Fundamental crisis in a philosophy of life? The environmental crisis is a crisis of identity. What the hell *are* we? What sort of decent relationship *can* we develop with this earth? What sort of relationship *must* we have?

I suppose some of you will think that I have been unnecessarily hard on Christianity, but I think not, because I am convinced that such religious ideas and all other superstitions and ignorance are the principal obstacles to understanding ourselves.

Perhaps our psychological and social attempts to make over the world in our own image is an extension of the old biological necessity to incorporate environment into organisms' image; if so, *we will only stop it when we have a decent ecological and evolutionary image of ourselves.*

You are an animal.

You get one run through.

The way to oneness is to live it.

> All this is perfectly distinct to an observant eye,
> and yet could easily pass unnoticed by most.
>
> **H. D. THOREAU**

NICE THINGS TO READ

Evolution

Berrill, N.J., *Man's Emerging Mind*. New York: Fawcett-Premier, 1955.

Boulding, K., *The Image*. Ann Arbor: University of Michigan Press, 1961.

Carrington, R., *Mermaids and Mastodons*. London: Chatto and Windus, 1957.

Darwin, Charles, *On the Origin of Species*. Cambridge: Harvard University Press, 1964.

DeBeer, Gavin, *Charles Darwin*. New York: Doubleday-Anchor, 1965.

Dunn, L.C., *Heredity and Evolution in Human Populations*. New York: Atheneum Publishers, 1965.

Eiseley, L., *Darwin's Century*. New York: Doubleday-Anchor, 1961.

———, *The Immense Journey*. New York: Ramdom-Vintage, 1957.

Greene, J.C., *Darwin and the Modern World View*. New York: New American Library-Mentor, 1963.

Hutchinson, G.E., *The Ecological Theatre and the Evolutionary Play*. New Haven: Yale University Press, 1965.

Keosin, J., *The Origin of Life*. New York: Reinhold Book Co., Inc., 1968.

Simpson, G.G., *The Meaning of Evolution*. New York: Bantam Books, Inc., 1971.

———, *Biology and Man*. New York: Harcourt, Brace, and World, Inc., 1969.

Smith, H.W., *From Fish to Philosopher*. New York: Doubleday-Anchor, 1961.

———, *Man and His Gods*. New York: Grosset & Dunlap Inc., 1956.

Williams, G.C., *Adaptation and Natural Selection*. Princeton, N.J.: Princeton University Press, 1966.

Ecology

Bleibtreu, J., *The Parable of the Beast*. New York: Macmillan-Collier, 1969.

Cain, A.J., *Animal Species and Their Evolution*. New York: Harper-Torchbooks, 1960.

Leopold, Aldo, *Round River*. New York: Oxford University Press, 1953.

Peattie, D.C., *Flowering Earth*. New York: Viking-Compass, 1965.

Animism

Hamilton, Walker, *All the Little Animals*. New York: Ballantine Books, Inc., 1969.

Leopold, Aldo, *A Sand County Almanac*. New York: Oxford University Press, 1949.

Lily, J.C., *The Mind of the Dolphin*. New York: Avon Books, 1969.

Lord, Russell, *The Care of the Earth*. New York: New American Library-Mentor, 1963.

Lorenz, Konrad, *King Solomon's Ring*. New York: Crowell-Apollo, 1952.

Mowat, Farley, *Never Cry Wolf*. Boston: Atlantic-Little, Brown and Co., 1963.

Sauer, Carl, *Land and Life*. Berkeley, California: University of California Press, 1967.

Tinbergen, Niko, *The Herring Gull's World*. New York: Doubleday-Anchor, 1967.

Humanism

Burgess, A., *The Wanting Seed*. New York: Ballantine Books, Inc., 1964.

Cooper, David, *The Death of the Family*. New York: Random House, Inc., 1970.

Eiseley, L., *The Firmament of Time*. New York: Atheneum Publishers, 1967.

Hall, E.T., *The Hidden Dimension*. New York: Doubleday-Anchor, 1969.

Hardin, Garrett, *Population, Evolution, Birth Control*. San Francisco: W.H. Freeman and Co. 1969.

Laing, R.D., *The Politics of Experience*. New York: Ballantine Books, Inc., 1968.

Naturalism

Anonymous "Four Changes," *The Environmental Handbook*. New York: Ballantine Books, Inc., 1970.

Castaneda, Carlos, *The Teachings of Don Juan*. New York: Ballantine Books, Inc., 1969.

Muir, John, *The Story of My Boyhood and Youth*. Madison, Wisc.: University of Wisconsin Press, 1965.

Shepard, Odell (Ed.). *The Heart of Thoreau's Journals*. New York: Dover Publications, Inc., 1961.

Snyder, Gary, *Earth Household*. New York: New Directions Publishers, 1969.

Thoreau, H.D., *Walden*. New York: Modern Library, 1950.

Veblen, Thorstein, *The Theory of the Leisure Class*. New York: New American Library-Mentor, 1953.

INDEX

Abstractions, 22
Acquired characteristics, 54, 63
Adaptation, 5, 8, 10, 15, 51, 53, 54, 74
Adaptive stance, 14
Adaptive systems, 15, 16
Aggressiveness, 46
Alternative reality, 81, 102, 104
American imperialism, 101
Animal behavior, 45
Animal empathy, 33, 34, 110
Animal sound, 34
Animism, 46, 105
Associations of atoms, 20
Associations of plants and animals, 11
Atomic potentialities, 3
Atoms, 20, 36, 37, 38
Aye-aye, 8

Balance of nature, 11
Beauty, 24, 29, 92, 101
Biological organization, 38
Biologists, 22, 85
Biosphere, 21, 81
Bird plumages, 10
Birth control, 60
Birth rate, 60

Bonding capacities, 13
Brain, 12

Capitalism, 82
Carrying capacity, 60
Catastrophism, 4
Cats, 20, 43
Cellular chemistry, 36
Chemical soul, 92
Chemical soup, 3
Christian theology, 13
Christianity, 36
Codes of information, 5
Color vision, 10
Communication, 63, 85
Community organization and metabolism, 24
Complexity, 3, 23, 24, 52
Context, 43, 51, 52, 56, 66
Control, 61
Convenience, 28
Corruptness of man, 16
Creationism, 4
Creator, 5, 10, 25
Cultural adaptation, 57, 63
Cultural evolution, 54, 55, 56
Cultural situations, 46, 53, 54, 57

Dame Nature, 24
Darwin, 4, 14, 23
Death rate, 60
Developmental processes, 6, 51
Devils, 14
Differential reproductive success, 5
Dignity, 16, 103
Direction of evolution, 4, 11
Diversity, 5, 11, 14, 44, 85, 86

Earth animals, 39
Earth awareness, 3
Earth child, 3
Earth elements, 20
Earth mother, 61
Earth stuff, 16, 38
Ecological context, 15, 66
Ecological criminal, 102
Ecological responsibility, 101
Ecologists, 23
Ecology, 23, 43, 88
Economic systems, 82
Ecosystem, 11, 80
Ecstasy, 90, 93
Ego control, 65
Elementary particles, 23
Emotions, 13
Empathy with animals, 33, 43, 110
Enlightenment, 67
Entities, 22, 36, 51, 65
Environment, 13, 15, 20, 22, 25, 36, 51, 53, 56, 61, 74, 83, 84, 90, 101, 104
Environmental biology, 23
Environmental deterioration, 44
Environmental problems, 27, 28, 57, 65, 74, 87, 93, 110
Environmentalists, 77
Enzymatic systems, 15
Equilibrium relationships, 57, 60, 61, 104
Eternity, 83
Ethical systems, 15
Ethics, 13, 15, 84, 95, 101, 108
Evolution of matter, 13
Evolutionary biologists, 23, 25
Evolutionary context, 15, 66
Evolutionary explorations, 3
Evolutionary origins, 15, 69
Evolutionary pathways, 22
Evolutionary perspective, 6, 11, 13, 16, 43, 69
Evolutionary process, 12, 15, 47, 74
Experience, 12, 13, 24, 44, 107
Extinction, 15

Flora, 29
Flowers, 10

Genepools, 4, 9, 22, 25, 51, 56, 58
Generational magic, 62
Genes, 9, 53, 58
Genetic message, 4
Genetic programming, 16
Genetic recombination, 13, 58
Genetic selection, 54
Genotype, 13, 51, 53, 56
Gods, 14
Goose-bumps, 10
Great chain of being, 10
Gregariousness, 66

Harmony of the living universe, 5
Hereditary material, 4
Historicity of the evolutionary process, 6, 14
Homesteaders, 27
Human animal, 45
Human ecology, 55

Identification with social group, 65
Images, 6, 9, 20, 111
Imagination, 55
Immortality, 13
Inalienable rights, 60, 62
Inexhaustible content, 104
Insect pollinators, 10
Insurance, 75
Intelligent non-manipulation, 55
Introgression, 25
Invidious comparisons, 75

Language, 12
Life-stuff, 38
Life style, 22
Life-support systems, 55, 84, 105
Linneaus, 25
Living organization, 38

Madness, 94, 102, 107
Materialism, 94, 106
Meaning, 15, 24, 43, 51, 53, 59, 62, 65, 66, 69, 74
Memory, 13
Metaphysical propositions, 12
Migration, 29
Mimicry, 10, 11
Mind, 12, 55
Mind-body problem, 12

Models of natural systems, 23
Molecular biology, 5
Molecules, 3, 12, 37, 38
Morality, 61
Mutations, 4, 58

Natural community, 11
Natural forces, 6
Natural selection, 4, 5, 10
Naturalistic ethic, 15, 109
Neighborhood, 27
Neurons, 21
Neurophysiologists, 12
Nodes, 20, 22
Noosphere, 21

Old Bitch, 25
Oneness, 111
Order, 24
Organism-environment dichotomy, 20
Origin of life, 3
Origin of species, 4

Peopling rocks, 14
Perceptions, 13
Personality, 13, 46
Philosophy, 46
Phylogenetic memory, 51
Physicists, 23
Pollution, 100
Populations, 4, 56, 58
Poverty, 103
Prairie, 29
Precambrian, 6
Predation by wasps, 7
Primal lover, 29
Primate ancestors, 10
Primate evolution, 9
Procreant urge, 11
Progress, 82, 94
Psychic wholeness, 66
Psychologists, 12
Purpose, 10, 11, 13, 15, 59, 101

Reality, 69, 73, 80
Receptiveness, 67

Recycling, 28, 38
Religious ideas, 14, 15, 56, 61
Replication, 3
Reproduction, 3, 5, 51, 52, 53, 56, 60, 62, 106
Responsibility, 69
Revolution of consciousness, 78
Rhubarb wine, 28

Sanity, 69, 87
Schizophrenia, 87
Science, 26
Scientific materialism, 3
Scientific natural history, 23
Scientific revolution, 12
Security, 27, 75, 79, 85, 99
Selective forces, 12
Selective pressure, 9
Selective process, 14, 51, 52, 53, 56, 59, 62, 106
Serenity, 65
Sexual recombination, 13, 58
Sexuality, 10, 59, 62, 66
Social behavior, 12
Social evolution, 56, 74
Social organism, 76
Social stability, 83
Soils, 38
Soul, 13
Speciation, 25
Species, 25
Spontaneous chemical synthesis, 3
Superorganism, 11
Survival, 10
Symbiosis, 11
Systematizing, 24, 25
Systems analysis, 26

Unexpected content, 44
Universe, 21

Variability, 58

Wasps, 7

Zoogeography, 22